THE STORY OF LIFE

THE STORY OF LIFE

T. R. E. SOUTHWOOD

OXFORD
UNIVERSITY PRESS

Great Clarendon Street, Oxford OX2 6DP

Oxford University Press is a department of the University of Oxford.
It furthers the University's objective of excellence in research,
scholarship, and education by publishing worldwide in

Oxford New York

Auckland Bangkok Buenos Aires Cape Town Chennai
Dar es Salaam Delhi Hong Kong Istanbul Karachi Kolkata
Kuala Lumpur Madrid Melbourne Mexico City Mumbai Nairobi
São Paulo Shanghai Singapore Taipei Tokyo Toronto

Oxford is a registered trade mark of Oxford University Press
in the UK and in certain other countries

Published in the United States
by Oxford University Press Inc., New York

First published in hardback 2003
Paperback 2003
Reprinted (with corrections) 2004, 2005, 2007

British Library Cataloguing in Publication Data
Data available

Library of Congress Cataloging in Publication Data
Data available

ISBN 978-0-19-860786-1

10 9 8 7

Designed and typeset by Pete Russell, Faringdon, Oxon
Printed in Great Britain by the
MPG Books Group, Bodmin and King's Lynn

Preface

THE aim of this book is to chronicle the history of life from the origin of the earth until the present day—with a glimpse into the future. It provides answers to questions such as how did life begin? What may have happened during snowball earth? Why did the dinosaurs become extinct? Are we all descended from an 'African Eve'? Will our activities change the pattern of life on earth to mirror a major extinction? It is intended for the lay person as well as the student.

The book's core derives from a course of lectures I gave for 18 years to first year life science students at Oxford. Their response, and that of some non-biologists who attended the lectures, has encouraged me to write this account.

An understanding of the story of life provides a benchmark for judgements on the environmental problems of today. The sea level has changed, as has the concentration of carbon dioxide and most other physical conditions. While life has survived, individual species and indeed whole communities of animals and plants have been wiped out. All these earlier extinction events have their origins in some physical change. Today the concentrations of carbon dioxide and some other gases are rising, but unlike earlier events, these changes have their origin in the actions of an organism—*Homo sapiens*. Our civilisation depends on very constant conditions and seen against this background we can see that we perturb our environment at our peril.

In covering this field I have been greatly helped by many colleagues. In particular Per Ahlberg, Paul Barrett, Tom Cavalier-Smith, Simon Conway Morris, Alan Cooper, Clive Hambler, Tom Kemp, W. J. (Jim) Kennedy, Stephen Moorbath, Paul Seldon, Adrian Thomas and Eleanor Weston have read various sections and David Raubenheimer has read the whole manuscript, as have, from lay veiwpoints, Charles Southwood and Belinda Wood. I am most grateful to them all as I am to Rosanna Wellesley and Jim Robins for the care with which they have executed the illustrations and to Cathy

Kennedy, of the Oxford University Press, for many positive contributions. I am also indebted to the Leverhulme Trust for an emeritus fellowship. Warm thanks are due to my wife for her continual support.

Oxford
January 2002

RICHARD SOUTHWOOD

Contents

Introduction 1

CHAPTER 1

Special Chemistry 7

Pre- and Early Archaean: 4550–3500 Mya

CHAPTER 2

A Quick Start 14

Middle and Late Archaean: 3500–2500 Mya

CHAPTER 3

Cells with Nuclei 25

Proterozoic: 2500–600 Mya

CHAPTER 4

Jellyfish, Polyps and Worms? 36

Late Proterozoic: 700–545 Mya

CHAPTER 5

Claws, Rasps and Shells 43

Cambrian–Ordovician: 545–438 Mya

CHAPTER 6

Sand, Mud and Shallow Seas 65

Silurian–Devonian: 438–362 Mya

CHAPTER 7

The Giant Continent Forms 89

Carboniferous–Permian: 362–248 Mya

CHAPTER 8

A Sparse Start 118

Triassic: 248–206 Mya

CHAPTER 9

The World of the Dinosaurs 136

Jurassic–Cretaceous: 206–65 Mya

CHAPTER 10

Today's World Dawns 176

Tertiary–Quaternary: 65 Mya–Today

CHAPTER 11

The Evolution of the Able Ape 213

20 Mya–30,000 Years Ago

CHAPTER 12

Humans: The Great Modifiers 233

40,000 Years Ago, Today and Tomorrow

Further Reading 257

Acknowledgements 259

Index 261

Introduction

CONSIDER the amazing variety of life today: the great herds of animals that roam the African plains, the shoals of fish that teem in coral reefs or the flocks of penguins that huddle on the Antarctic ice. Yet what we see around us is but one still from the film of life, a glimpse that we can only understand if we know what came before. This is the book of the film of all life from the beginnings of the earth. There are a number of different clues to the earlier parts of the film. To follow these, one needs to understand how the physical environment influences life, the process of change (evolution), how one form of life impacts on others and how living things are classified.

The film begins some 4550 million years ago (Mya) with the birth of the solar system and the formation of the earth. Until 3900 Mya the earth was so heavily bombarded with asteroids from space that life could not exist. The first probable signs of life occur in rocks laid down 150 million years later. So much special chemistry must have taken place over this period that some people have suggested that life must have come from space. It probably did not, but some essential chemical components probably did, giving it a quick start.

The story of life is not a steadily unfolding pattern. It is as if we are looking at a kaleidoscope and every now and then it is given a shake so that some components of the picture disappear, others remain, some are altered and some new ones appear. The 'shakes' are due to changes in the physical environment, such as a collision with an asteroid or climate change which can lead to the fall or rise of sea level or the spread and retreat of ice sheets. As the tectonic plates that make up the crust of the earth move around, the earth's geography has changed, and it is still changing continuously (the North Atlantic is widening by about a centimetre a year). Areas that had a tropical climate may move towards the poles. Sometimes there has been land at the South Pole, sometimes only ocean, with huge implications for the world climate. Besides geography, another factor that influences climate is the amount of 'greenhouse gases', especially carbon dioxide, in the air. At one point in time the whole earth seems to have been like a 'snowball', with ice at the equator.

Life is special because it is not a passive participant in events. The very variety of life is what enables the process of change, of evolution, to occur. As some individuals die and others live to reproduce, there is natural selection. That is what happened to the very first living organisms, in the 'primordial soup', and this is what is happening to lions and mice, oak trees and jellyfish today. The concept of natural selection was first put forward by Charles Darwin and Alfred Wallace in 1858 and it is significant that they developed it, independently, after observing the diversity of life on their extensive travels. Biologists gain inspiration by observing and identifying patterns.

The basic concept of evolution is simple. Some organisms have more surviving offspring than others that differ from them in some way (and are said to be less 'well adapted'). However the fundamental characteristics of organisms are determined by the exact form of the DNA of their genes which can lead to unexpected evolutionary outcomes. How probable it is for an individual to die in a particular circumstance is likely to depend on its qualities, such as—in animals—how it behaves. For example, whether it runs away or fights. Now it may be against the individual's advantage to stand and fight—it may die in the process—but to the advantage of other individuals who can avoid fighting as a result. If at least two of those other individuals are of the same genetic makeup as the fighter, more of its genes will survive into the next generation because the individual died in the fight rather than saving its skin by running away. This is exactly what happens in ants and certain aphids, where relatives have a virtually identical genetic makeup, and is the basis of Oxford biologist Richard Dawkins' 'selfish gene' concept. One can think of evolution being the product of the genes which are selfish, each seeking to have as many replicates of itself as possible in the following generation. The organism is the carrier of this parcel of genes and, in the case of these ants and aphids, the genes prosper because some individuals that carry them do not survive.

Ants and aphids have unusual inheritance mechanisms. In most organisms, other than bacteria, the individuals are unique so that the genes are primarily 'concerned' with the survival of the individual that is carrying them. Thus, in general, when an individual possesses some slight genetic difference that confers an advantage over its peers, it will have more offspring that survive, and in succeeding generations the gene responsible for the advantage will spread through the population. Sometimes the environment changes so that a gene that was rare gives an advantage and spreads.

The development of resistance to antibiotics or to insecticides are examples of this. Suddenly, a new chemical in the environment kills most of the population, but among the variety of the population's genetic make-up there are one or more individuals that survive and reproduce because they have an unusual biochemistry that prevents the chemical from killing them. Next time the chemical is applied they comprise a larger proportion of the population and if the treatment is maintained this process of selection continues until the entire population is completely resistant; all individuals carry the gene that provides the special biochemistry. However, the ability to survive this poison is unlikely to be the only effect of that particular genetic make-up; under normal conditions it may well carry a handicap of some kind so that once the treatment is stopped the resistant gene will become rare again. So resistance is avoided and this is why pesticides and antibiotics should not be applied continuously—to do so brings the danger of resistance.

Evolution can therefore be seen as the selection of different gene packages—usually comprising the individuals that contain the genes to suit different environments. If the individuals no longer meet and therefore can not pair (technically gene flow ceases) two new species can come into being. The commonest barriers to meeting are geographical—mountain ranges, valleys (for mountain species), seas, or deserts. Less frequently speciation may occur in the same general location. In this case the separation may be in time—mating being at different seasons—or it may be due to their being restricted to different microhabitats.

Once two species are separated, gene flow ceases, but their DNA make-up will continue to diverge. Work in the last few decades has shown that random changes occur more or less continuously in the detailed structure of the DNA. The majority of these changes are neutral, that is they do not make the individual more or less well adapted or 'fit', but they spread through the population and contribute to a genetic fingerprint characteristic of the species. Since the rate at which such changes accumulate is believed constant and is roughly known, differences in DNA between different species will reflect the time since they last regularly interbred; that is the time since they were but one species. The greater the difference, the more random the changes and hence the longer the time. This is the basis of the 'molecular clock' that measures the time since separation of two species (see Chapter 2, Fig.2.1); in terms of the metaphor, it shows the length of film.

Animals, plants and microbes are studied across the world and scientists need to be sure they are talking about the same species of organism. Each species therefore is given a distinct scientific name that consists of two parts, for example, people are *Homo sapiens*, dogs *Canis canis* and daisies *Bellis perennis*. These scientific names are always printed in italics. They are a tool, like a microscope or a test-tube. The first word is the name of the genus that includes other closely allied species (e.g. Neanderthal Man, *Homo neanderthalensis*) and, by convention, is always written with a capital. The second word is the unique name for the species and commences with a small letter. This binary system of nomenclature, as it is termed, was introduced by the Swede Carl Linnaeus in 1758 and is used throughout the world, enabling scientists to exchange information knowing they are talking about the same species—provided of course they have identified it correctly! When a new species is discovered, which happens frequently with insects and now rarely with birds, it is formally described and, with its newly invented name, this description is published in a scientific journal.

The many species of organism are classified. The most practical method, and the most longstanding, is known as the 'biological system'. It involves grouping organisms in a hierarchy of categories (taxa); all those in one taxon are considered to be more closely related in evolutionary history to each other than to any member of another group of the same rank. A whole series of ranks are used, but those most widely adopted can be applied in the classification of the common housefly:

kingdom—Animalia
phylum—Arthropoda
class—Insecta
order—Diptera
superfamily—Muscoidea
family—Muscidae
subfamily—Muscinae
genus—*Musca*
species—*domestica* Linnaeus.

The name Linnaeus at the end indicates that it was Linnaeus that originally described it and if one is uncertain as to what he meant (imagine two different house flies are found, both of which fit his description) one goes to look at his original specimen known as the type. Hence the great importance in museum collections for maintaining the international and comprehensive system of nomenclature.

The disadvantage of this system is that there is no really objective way of allocating a group of species to any specific level, say, a family or a subfamily. The species involved will all have certain features in common which the scientist working on the group will use to define a given level. But which characters are included is subjective. For this reason the German entomologist W. Hennig proposed in 1950 another form of classification, termed 'cladistics'. Cladistic classification is based on recognizing the sequence in time in which characters evolved. When it is recognized that a new character has evolved in a group, then all the species with this character are said to belong to a 'clade'. Within this clade another character may well have evolved and only a few species will have come from that stock; they form another clade, nested within the first clade. Thus a clade has no particular hierarchical level. Cladistics has complex rules about how defining characters are chosen and an even more complex terminology. A cladistic analysis of a group tells us a great deal about its evolution, but for descriptive purposes the old biological system is more practical. For example, cladistic analysis shows clearly that birds evolved from a group of reptiles, and as they have all the basic characters of reptiles they should strictly be described as members of the reptile clade! The approach I have taken in this book is to adopt cladistic interpretations of evolution, but to use the biological system for the descriptors of groups of organisms.

CHAPTER 1

Special Chemistry

Pre- and Early Archaean 4550–3500 Mya

THE solar system was born some 4550 million years ago when a star exploded at the end of its life. This supernova scattered matter and energy into a cloud of interstellar dust and gas and as a result most of the material came together to form yet another star, our sun. Various other chunks of material collided and coalesced and began to form the planets. The gravitational pull of the nascent earth would have attracted other pieces of debris. Many of these were large—and one was so large that its impact tilted the earth on its axis and knocked off chunks that came together to form the moon. These impacts would have generated great heat, probably enough to melt the earth. Another, and continuing, source of heat, is that generated when radioactive elements, like uranium, decay. Which source of heat was most important is uncertain, but together they were sufficient to allow the heavy materials, principally iron, to flow to the earth's centre, where they exist as a molten core. This core is now nearly three and a half times more dense than the outer layer, the crust, and just over twice as dense as the middle layer, the mantle. Both the crust and mantle are solid. Compared with the size of the earth the crust is very thin; it has been likened to a layer of paper stuck on the outside of a football.

When the minerals of the earth were heated to these very high temperatures various gases would have been given off; this still occurs in volcanic eruptions. The gravitational field of the earth, unlike that of the moon, is sufficient to hold them as an outer layer—the atmosphere. The nature of these gases can be found by studying those given off by volcanoes and those produced when minerals from deep in the earth are heated to very high temperatures. These investigations show that the major constituents of this early atmosphere would have been nitrogen, water vapour and carbon dioxide. Smaller amounts of methane, ammonia, carbon monoxide, sulphur dioxide, hydrogen sulphide and hydrogen cyanide were probably present, some arising from the effects of lightning, asteroid impact and ultraviolet radiation. Comets are thought to be an alternative source for some of these gases, particularly water vapour. The water vapour would have condensed and fallen as rain (with some of the other gases dissolved in it) to form the earliest oceans. It is likely that the impacts of asteroids would have caused the early oceans to evaporate only to reform by condensation. The early atmosphere lacked oxygen; if there were trace amounts these would have quickly combined with some of the large quantity of iron still present in the oceans and in the surface layers of the land.

When rocks are formed from the liquid core of the earth various radioactive elements, especially uranium and thorium, are trapped in the crystals. These elements decay very slowly; each atom of an isotope[1] of uranium or thorium eventually becomes an inactive isotope of lead. Ancient rocks are dated by the ratio of the lead and uranium (or thorium) isotopes. The slow rate of decay means that long periods of time must pass before half the uranium atoms have been transformed into lead, this time is known as the half-life. One uranium isotope (^{238}U) has a half-life of 4500 million years; after this period half the original amount of uranium 238 would have transformed into the stable lead isotope (^{206}Pb). A 'sealed sample' of the uranium and the derived lead is found in crystals of zircon, a mineral formed in molten rock.

This method, using the ratios of uranium and lead isotopes, cannot be used to determine the age of the earth itself as none of the original rock is now on the surface. The crust of the earth consists of tectonic plates, seven large ones and about 20 smaller ones. These move around the surface of the earth and where they meet on the floors of oceans one or both may turn under towards the centre of the earth. As a result the surface of the earth exposed in the first thousand million years of its life is now lost; it has been drawn back into the interior. The original surface of the moon is, however, still exposed and the many large impact craters there show that the earth, being in the same region of space, was similarly bombarded in its early life; this is believed mainly to have occurred around 3900 Mya.

However all the solid material of the solar system, such as the earth, the moon and small lumps that crash on the earth as meteorites, was 'born' at the same time. Therefore the relative abundance of the four isotopes of lead

[1] An isotope is one of two or more forms of the same element that differ in their atomic weight; this is due to differing numbers of neutrons—particles that have no electric charge—in the nucleus of the atom. Isotopes are designated by their atomic weight followed by the symbol for the chemical element.

(known as ^{204}Pb, ^{206}Pb, ^{207}Pb and ^{208}Pb) in a meteorite, which does not contain thorium or uranium, will be that of the primordial earth. On the earth today radioactive decay of thorium and uranium will have increased the proportions of ^{206}Pb, ^{207}Pb and ^{208}Pb; as the rates at which the radioactive decay occurs are known the age of the earth can be determined by a comparison of the isotopic composition of lead in meteorites with the average of that in terrestrial lead.

The formation of organic molecules

The building blocks of living organisms are organic molecules: carbon atoms combined with those of oxygen, hydrogen and nitrogen, and frequently also with those of sulphur and phosphorus. They are produced by the actions of enzymes and other special chemistry in living cells, but this begs the question: how did this process start?

Charles Darwin recognized this problem, which was used as an argument against his theory of evolution. On 1 February 1871 he wrote to his friend, the botanist Joseph Hooker:

> It is often said that all of the conditions for the first production of a living organism are now present, which could ever have been present. But if (and oh! what a big if!) we could conceive in some warm pond, with all sorts of ammonia and phosphoric salts, light, heat, electricity etc. present, that a protein compound was chemically formed ready to undergo still more complex changes, at the present day such matter would be instantly devoured or absorbed, which would not have been the case before living organisms evolved.

Darwin visualized a 'primordial soup' of complex organic materials and was almost certainly right in his view that today living organisms would consume these particles. But the environment of the early earth was far from the benign picture conjured up by the 'warm little pond'; it was dominated by ferocious volcanic activity and pounded by asteroids, meteorites and comets, as well as by cosmic rays and ultraviolet radiation. The absence of oxygen meant that there was no ozone screen to reduce this ultraviolet radiation and the radioactivity from elements in the earth's crust would have been about two to three times what it is today.

The next major contribution to understanding the origin of life was made by the Russian botanist Aleksandr Oparin. In a small book published in 1924 he proposed that the steps towards life would have been from inorganic substances to more complex organic ones (i.e. the compounds of carbon) and from these to protocells, and so to living organisms. He also suggested that the first organisms were most likely to be heterotrophs; that is they would have obtained their energy by feeding on organic materials, in this case the complex molecules in the primordial soup. This way of feeding contrasts with that of other organisms, like most plants, which usually obtain their energy from sunlight, or certain bacteria that perform the same function by simple chemical processes; such organisms are known as autotrophs—self feeders.

It was not until 1953 that this concept of life's origin was further advanced when Stanley L. Miller and Harold C. Urey published a paper in which they described how simple organic molecules could be produced experimentally in conditions and from ingredients that they thought similar to those of the early earth. The experiment involved passing a strong electric spark, simulating lightning, through an atmosphere of hydrogen, methane, ammonia and water vapour. After a week (the equivalent, Miller and Urey thought, to millions of years of lightning), they found that several amino acids had been formed; three of these were known components of the proteins of living organisms. They had confirmed Oparin's suggestion that simple organic molecules could be formed only in the absence of oxygen, in what is termed a 'reducing atmosphere'.

However on the early earth impacts of meteorites, volcanoes, cosmic rays and solar flares were likely to have been the main sources of energy, rather than lightning, and the atmosphere would not have been rich in hydrogen, ammonia and methane. Many more experiments have been done with different mixtures and sources of energy other than electrical discharges. These experiments have shown that if energy is applied to solutions with mixtures of various inorganic compounds containing the elements carbon, nitrogen, oxygen and hydrogen then a wide range of simple organic compounds (amino acids, fatty acids and sugars, such as glucose) is produced. If free oxygen is present no organic compounds are formed, so ironically the oxygen-free atmosphere of the early earth was not an obstacle to life, as might be supposed from its role today, but a necessary condition. The solution needs to be slightly alkaline; if it is acidic the compounds produced are not those involved in living processes.

A different scenario was proposed a few years ago by the German scientist, Günter Wächtershäuser, who suggested that the earliest organisms were not heterotrophs, as proposed by Oparin , but chemoautotrophs which obtained their energy from the breakdown of chemicals. He postulated that their energy source was provided by carbon dioxide or carbon monoxide being brought together with hydrogen on the surface of iron pyrites, found widely in the earth. The conditions that he considered suitable for this are still found around hydrothermal vents (see page 33) in the deep oceans. So far no one has been able to carry out experiments like those of Miller and Urey to demonstrate this process.

It has often been suggested that life arrived on earth from space, but most scientists are doubtful if any organism could survive the journey through space and the atmosphere. But organic compounds could survive and may have come with comets, meteorites or galactic dust. In the last decades two pieces of evidence have confirmed that organic molecules do form in space. Some meteorites, termed 'carbonaceous chondrites', contain coal-like material. However such organic material could have resulted from contamination between the time the meteorite landed and when it was analysed. In September 1969 a chondrite landed near Murchison, in Australia, and pieces were quickly retrieved and carefully analysed. Numerous amino acids and other organic molecules were found, their proportions indicating that conditions rather like those envisaged by Miller and Urey were present on the asteroid from which the meteorite had come and that extraterrestrial organic molecules are real.

The second type of evidence for the existence of organic molecules in space is derived from the analysis of the spectrum of the light from interstellar gas and dust. Certain amino acids, such as glycine, have been found using this technique, while some research suggests that more complex molecules and polymers such as cellulose may also be present in galactic dust.

It has been calculated that 10–15 per cent of galactic dust particles and comet debris is organic, which leads to the calculation that several thousand tonnes of organic molecules settle on the earth each year. Undoubtedly this 'galactic rain' has contributed to the accumulation of organic molecules on earth, supplementing that produced on the earth by the various forms of energy with which it is bombarded, such as ultraviolet and electrical, as well as that produced from the impact of asteroids and meteorites. Further major energy source is the internal energy released through volcanoes.

From molecules to simple cells

Whereas there is now hard scientific evidence about how simple organic molecules have been formed, the transition from molecules to simple cells is largely a matter of informed speculation. At least five steps can be recognized:

1. concentration of the biologically important simple molecules ('monomers');
2. the joining together of a series of monomers to form biological polymers (examples are starch, collagen and cellulose);
3. formation of an outer membrane to provide a microenvironment where the special chemistry characteristic of life could occur;
4. development of a mechanism to provide energy;
5. information transfer to permit cell replication.

A number of sites have been suggested where molecules could be concentrated. When seas abut deserts or the laval flows from volcanoes, drying and concentration occurs. Large parts of the oceans are covered with foam and the bubbles would have provided enclosed space and so a potential site. On land clay particles are a candidate; they consist of sheets of atoms and the spaces between these sheets may have served as a template for the concentration and organisation of the simple molecules. Because the early earth was repeatedly struck by asteroids it might be argued that the only place where organic molecules could have survived these cataclysmic upheavals would be the deepest oceans; here vents from submarine volcanoes provide a hot environment rich in sulphur and other inorganic compounds. In these conditions, as Wächtershäuser proposed, the surface of iron pyrites could be the site of synthesis of organic molecules. Concentration could also have taken place in those parts of the solution under deep ice caps that do not freeze.

In living organisms virtually all polymers are formed by enzymatic action. If they are to be synthesized away from cells and without enzymes, the process requires special conditions (such as high pressure) which could have been present in some of the sites where it has been argued that the concentration of the simpler molecules could have occurred. Perhaps some other polymers arrived with galactic debris.

The membrane of living cells is important because it controls the chemicals that enter and leave; thus chemical reactions, especially those involving enzymes, can occur in the cell which would be impossible in more open surroundings. In certain mixtures of polymers and water, droplets form in which some enzymatic reactions are facilitated. Clay templates might also provide a 'microlaboratory', as too might protenoid microspheres. These are formed if amino acids are heated to relatively high temperatures (causing some polymers to form); the material is then added to water when spherical bodies with a double outer layer are produced. Hydrothermal vents and hot springs, in areas of volcanic activity, would provide the right conditions.

The basis of energy production in most living organisms is the reaction with water (hydrolysis) of adenosine triphosphate (ATP). One of the first evolutionary steps in this direction would have been the conversion of glucose (from the primordial soup) into ATP by a process called glycolysis. This biochemical pathway has remained basic for most organisms throughout evolution. Life would have ended when the glucose in the primordial soup was exhausted had not anaerobic photosynthesis evolved as it has in the cyanobacteria (see FIG 2.3). In this process energy from light is harnessed by a pigment called bacteriorhodopsin to produce ATP. In sulphur bacteria, hydrogen sulphide is the source of the hydrogen needed for ATP production in these organisms, and leads to sulphur deposition.

The next, vital step for the earth as we know it, was the development of photosynthesis, which produces oxygen. In this process water (H_2O), rather than H_2S is used as the source of hydrogen and so oxygen is liberated. Light is captured by various pigments, especially types of chlorophyll. The advent of oxygen allowed a further pathway for energy production to evolve; sulphur or other inorganic substances could combine with oxygen, a process known as 'burning'. The bacteria that obtain their energy from the reactions of inorganic chemicals in this way are known as chemoautotrophs.

A fundamental chemical step was the production of nucleic acids which control the production of other molecules of the same type (autocatalysis). This process has been demonstrated in the laboratory for a particular nucleic acid type compound. The key material for life is DNA, deoxyribonucleic acid, the famous double helix; it encodes the information for the functioning of the cell. This information, the instructions for the production of particular proteins, is carried by the messenger RNA, allowing the assembly of the proteins, the molecules that 'do things' in the cell. It seems likely that RNA evolved first, hence the expression, used by many scientists, 'an RNA world'. Initial RNA production may have arisen in hot acidic sulphur-laden waters rich in molecules consisting of chains of amino acid (peptides).

At one time it was believed that all these steps would have taken many hundreds of millions of years to complete. However new discoveries show that life seems to have started relatively quickly after the major bombardment with asteroids stopped about 3900 million years ago. Perhaps it may even have arisen before this time only to be extinguished by gigantic collisions and was subsequently restarted.

CHAPTER 2

A Quick Start

Middle and Late Archaean 3500–2500 Mya

THREE types of evidence indicate when life commenced and when certain types of organisms evolved. The first is based on fossils, the remains of organisms held in rocks where the actual parts of the organism have usually been replaced by a mineral. Less frequently the fossil records a track made by an animal—a worm's burrow or the footprints of a trilobite or a dinosaur. Sometimes the form of the whole organism is retained in the fossil, but more often only bits and pieces are preserved, and when these are first discovered they are often thought to represent different organisms and are given different names! The hardest parts, such as tree trunks, bones, and claws, retain their shape longest, and so are more likely to be recorded in the mud as it is compressed. Organisms that lack hard tissues and the soft parts of other organisms will quickly decay; it is only in exceptional circumstances, such as a massive mud slide, that fossilization is so quick that their form is captured.

The second type of evidence is provided by the ratios of the two stable isotopes of carbon: ^{12}C and ^{13}C. The atoms of the former are slightly lighter and are captured more frequently by the enzyme in the first step of photosynthesis—the basic step in the construction of the organic molecules used by most microbes and all plants and animals. Therefore, carbonaceous material derived from living systems (based on photosynthesis) will have relatively more atoms of the light isotope of carbon than are present in carbon dioxide. The opposite occurs when limestone is formed from the combination of carbon dioxide with calcium ions in sea water. A certain type of limestone provides a standard, against which any sample can be compared using mass spectroscopy. If the proportion of the heavier carbon isotope is less than that in the limestone this result is expressed as a negative value. Such values are the fingerprints of life.

The third type of evidence is derived from molecular biology and is currently an area of rapid progress. Evolution has occurred by changes in the genes and so the greater the differences in their genes the further apart two organisms are in their evolutionary history. Changes can be recognized by differences in the sequence of parts of the genes. We can compare the DNA sequences of genes for various basic proteins, such as the cytochromes (pigments fundamental to photosynthesis and respiration). Random changes in the sequence result in the substitution of one protein for another. These changes take place at a certain rate; the more changes the longer the time taken to generate them. These changes with time are described as the molecular clock for the group of organisms concerned. The clock has to be calibrated by some event whose time is known, generally from the fossil record. Having independently established the speed of the molecular clock, a phylogenetic tree can be developed (FIG. 2.1). Such a tree, indicating the last time when different organisms had a common ancestor, can be extended back in time beyond the date indicated by the fossil record by using sophisticated statistics and powerful computer programmes. As the rate of change of even random changes may vary, the validity of such calculations should

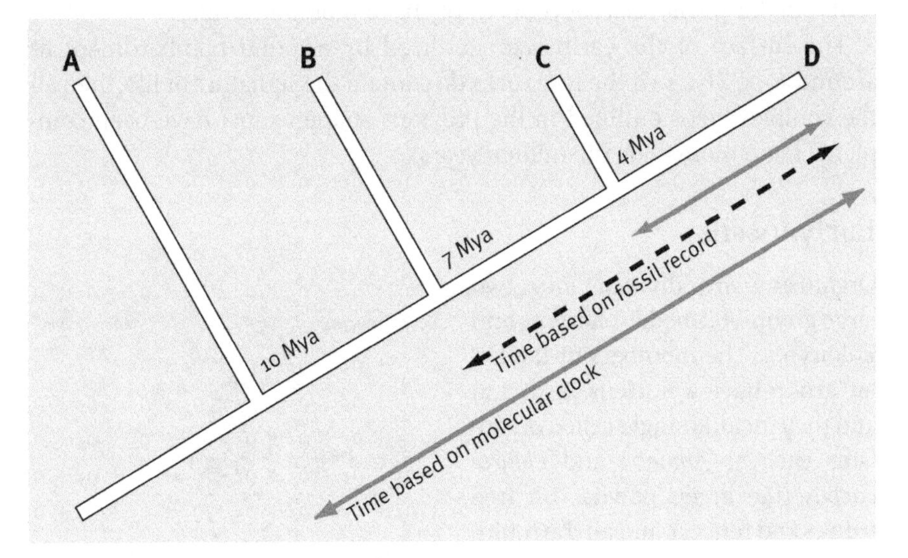

FIG. 2.1 A phylogenetic tree. The molecular clock is set by the time of the divergence of B from the common ancestor of C and D as known from the fossil record (7 Mya). Assuming that the clock's speed remains constant it is then used to determine the other times of divergence (10–4 Mya)

be based on several related lines and tested by comparison with estimates derived from changes in the sequence in one or more other genes. The genes used may be those in the nucleus or those in other cell bodies that contain DNA, such as the energy-producing mitochondria.

The oldest rocks

The oldest rocks that are still at the surface of the earth are the Isua formation in south-west Greenland and they date from 3750 million years ago. They are sedimentary and volcanic, both were laid down in shallow water, but have since undergone several periods of heating and compression. No fossils have been found in these rocks but they do contain carbon in the form of graphite. The proportion of the lighter isotope of carbon (^{12}C) is less than that characteristic of material that has been living, but higher than that for materials with a limestone origin. It has been argued that when subjected to high temperatures the carbon isotopes may redistribute themselves between carbonates and the remains of living material. So the proportion of the lighter isotope of carbon in the Isua rocks may be evidence of photosynthetic activity, of life. Alternatively it may represent the remains of the molecules in the 'primordial soup' or simply be of inorganic origin.

The surface of the earth was sterilized by asteroid bombardment at around 3900 Mya so if the Isua rocks do contain the signature of life, then all the complex steps outlined in the previous chapter must have been completed in no more than 150 million years.

Early fossils

Organisms are divided into two large groups, termed Prokaryota and Eukaryota. The members of the latter group have a nucleus (FIG. 2.2) and they include single-celled organisms such as *Amoeba* and *Pleurococcus* (the green powder on tree trunks and fences), and all the multicelled organisms we term animals and plants. The simplest and undoubtedly the first organisms are the prokaryotes that lack a nucleus

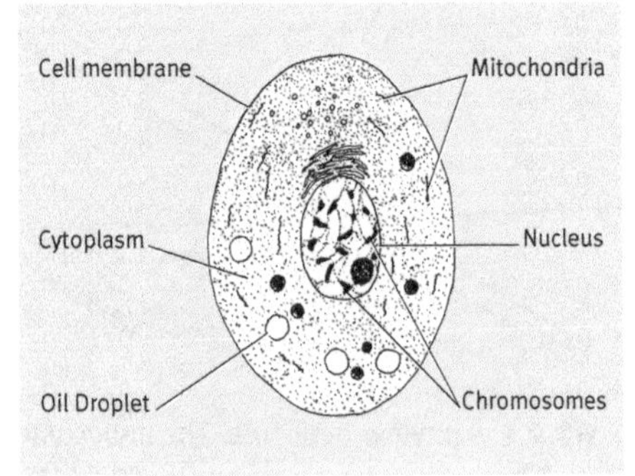

FIG. 2.2 A generalized animal cell, stained and showing chromosomes

and which may be loosely termed bacteria. This means that the earliest fossils will be those of bacteria—very small and simple. The identification of such fossils is difficult and was first achieved only in 1954 by Stanley A. Tyler and Elso Barghoorn (confirmed shortly afterwards by Preston Cloud), in the Gunflint Chert from southern Ontario. This rock dates from 2100 Mya.

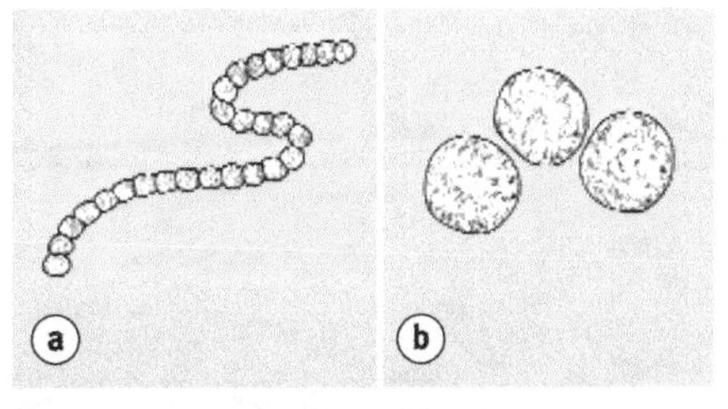

FIG. 2.3 Cyanobacteria; (a) *Nostoc*, (b) *Prochloron* (after Tudge 2000)

Since then even earlier fossils are thought to have been identified. The oldest come from the Pilbara Range in Western Australia, from rocks called the Warrawoona chert and the Apex chert which date from some 3500 Mya. Several different types of fossil have been recognized[1]; they are probably photosynthesizing bacteria, such as cyanobacteria that are still common as a thin layer of green slime on damp stones or where a tap drips into a basin (FIG.2.3). Perhaps surprisingly most of these ancient fossils can be related to forms occurring today. Indeed those from the younger (2800 Mya) Fortescue deposits (also in Western Australia) even seem to be fossils of species indistinguishable from cyanobacteria living today. This group shows a very slow rate of evolutionary change.

Just how slow is also shown by the presence today of structures termed stromatolites, fossilized examples of which are widespread in rocks as old as the Warrawoona chert. They are formed as small mounds in shallow water (FIG. 2.4), each one representing layer upon layer of mats, produced by bacterial activity. The uppermost layer consists of filamentous blue-green cyanobacteria; their photosynthesis, using the pigment chlorophyll, absorbs most of the light and produces oxygen. Below them are purple bacteria; these have a different pigment, bacteriochlorophyll, that absorbs wavelengths of light not trapped by chlorophyll, but their photosynthesis does not produce oxygen—it is anaerobic. Under this layer are other bacteria that can also live without oxygen and depend for their nourishment on dead bacteria from the layers above. The uppermost cyanobacteria produce slime and this traps minute particles of limestone and other minerals; eventually this layer screens out so much light that the bacteria glide through it to form a new set of layers. In this way hundreds of paper-thin layers are formed and

[1] While this book has been in press some new research has thrown doubt on the interpretation of these objects as remains of living organisms (see *Nature* **417**, 782–784; 20 June 2002)

FIG. 2.4 Stromatolites in Shark Bay, Western Australia. (These hummocks are mostly in the range 50–100 cm in diameter and about 30 cm high). For over 1000 million years stromatolites represented the only impact of living organisms on the landscape

the resulting rock, sectioned and polished, makes a very attractive architectural feature. In China they are called 'flower ring rocks'.

Various forms of stromatolites, no doubt formed by different bacteria, are found in the geological record from 3500 Mya for nearly another 2000 million years. So for most of the time that life has existed on earth its dominant representatives have been bacteria (FIG. 2.5). We can get an idea what the landscape must have looked like from the few existing sites where stromatolites are being built today (FIG. 2.4).

The two main sites where this occurs are Shark Bay in Western Australia and along the Sinai shore of the Red Sea. In both these localities the stromatolites exist in very salty warm water. The water is so salty that animals such as sea urchins and various molluscs cannot live; if they were present they would graze down the cyanobacteria and so the whole process of forming

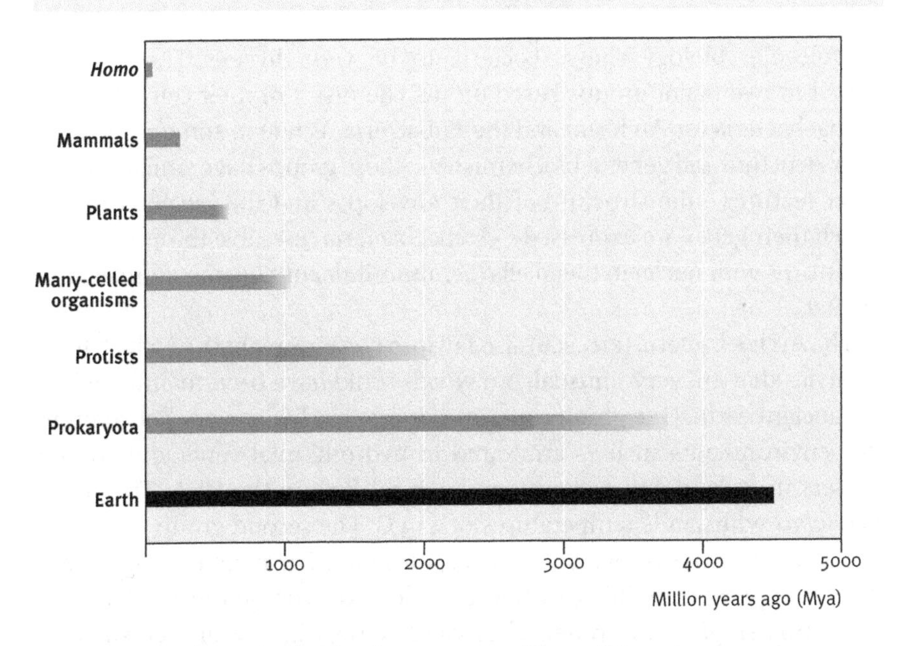

FIG. 2.5 Timescale showing the occupancy of the earth by various organisms (fading line represents uncertainty)

successive mats would stop. The evolution of these grazing animals has ended the widespread occurrence of living stromatolites.

A world of bacteria (Prokaryotes)

Bacteria were the only form of life on earth for much of its history (FIG. 2.5). They are still extremely numerous; it has been calculated that each of us harbours more individual bacteria than the total human population, past and present. Bacteria are still the only living organisms in many places that we term harsh or inhospitable, ranging from boiling volcanic springs to deep crevices in rocks and the Antarctic ice. Recently it has been claimed that living bacteria will develop from spores included in salt crystals in a geological formation laid down 250 million years ago! Furthermore many species living today are, like certain cyanobacteria (p. 17), close to, if not the same, as those active 2000 million years ago.

Bacteria multiply by simple division (one cell becomes two) or by budding (one cell grows out of another). In favourable conditions some species can divide every few minutes, but it has been suggested that those living in deep fissures in rocks may do so only once in 500 years.

Molecular biology shows bacteria to be very diverse. They can be placed in two major groups based on the chemistry of their cell walls: the Archaebacteria (or Archaea) and the Eubacteria. While in some respects—their structure and general biochemistry—these groups have similarities, in other features—the structure of their envelopes and the mechanisms by which their genes are expressed—Archaebacteria resemble the eukaryotes, organisms with nuclei in their cells (i.e., unicellular organisms, animals, and plants).

The Archaebacteria (FIG. 2.6) also fall into two groups, both of which have lifestyles that are very unusual, but which could have been maintained on the ancient earth. One group contains species that live in very hot sulphur-rich environments such as are found in hydrothermal vents and around geysers in Iceland and Yellowstone National Park in the USA. The genus *Pyrolobus* withstands temperatures of 113°C. The second group of archaebacteria have two distinct, but unusual, lifestyles. The methanogens obtain their energy by combining carbon dioxide and hydrogen to produce methane (marsh gas) and water. They can live only in oxygen-free environments such as lake and marine sediments, sewage farms, and the guts of certain animals. Are they survivors from an oxygen-free world? The halophiles live in concentrated salt solutions, as are found in desert lakes. One example is *Halobacter.* It photosynthesizes using bacteriorhodopsin, which gives it a purplish-pink colour; this type of photosynthesis may have preceded that based on chlorophyll.

There are many types of Eubacteria that would have been present in the world of bacteria which constituted life on the ancient earth: the purple bacteria and cyanobacteria that lived in stromatolites; others, like the Archaebacteria, which demanded the high temperatures and oxygen-free environments of hydrothermal vents. Many can thrive in anaerobic condi-

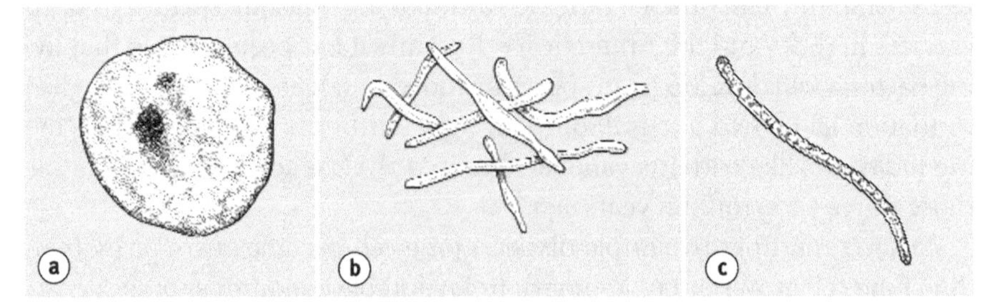

FIG. 2.6 Archaebacteria: (a) *Thermoplasma*; (b) *Halobacterium*; (c) *Sulfolobus* (after Tudge 2000)

tions and have unusual forms of metabolism; nowadays there are bacteria that live in and on plants and animals, some causing disease.

There are often strong links between the members of bacterial communities living together. Firstly, they may exchange genes, a process important in their evolution. Secondly, some bacteria living together in oxygen-free environments may form what is termed a consortium: one species uses a chemical for its life processes that is present in the environment. Sulphur frequently has this role. However, the waste product from this process is toxic to the species producing it, but it is the food of the second species in the consortium. The first species could not survive unless its waste was removed—there is a functional interdependence. There may be several species in a consortium. Bacteria also form consortia with more complex organisms.

Where did life start?

Life may have started in any of the sites where inorganic molecules could have been concentrated (p. 12). However some phylogenetic trees (FIG. 2.7) suggest that various Archaebacteria that live in hot environments have the deepest branches in the phylogenetic tree. These include species that live by metabolizing sulphur. So life may have evolved in such habitats—hydrothermal vents in the deep ocean or hot springs at the edges of

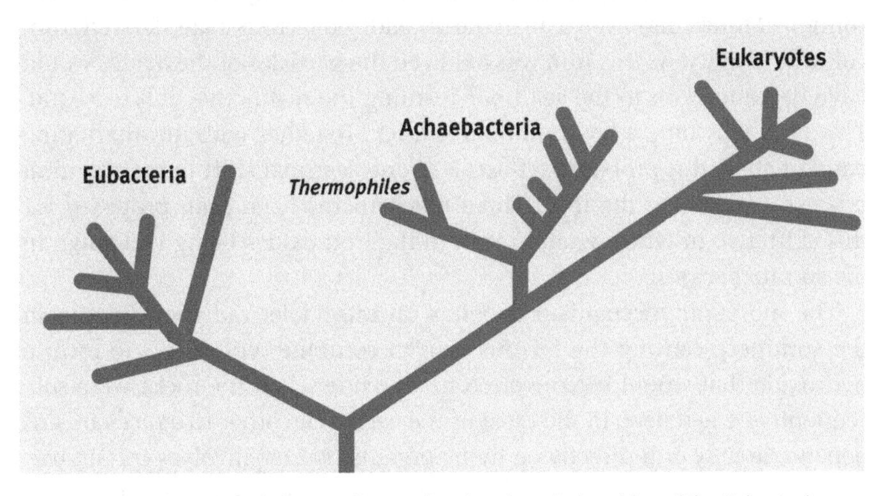

FIG. 2.7 A universal phylogenetic tree showing the relationships of the Eubacteria, Archaebacteria, and Eukaryotes based on rRNA sequence comparisons (after Woese *et al.*(1999)

volcanoes. If, as has been suggested, the earliest organisms obtained their energy by bringing carbon dioxide or carbon monoxide together with hydrogen on the surface of iron pyrites (p. 11) then these sulphur-rich environments would provide the appropriate conditions for such chemistry.

The alternative and more widely-held view, originally put forward by Oparin (p. 9), is that the first organisms lived on organic compounds such as glucose in the primordial soup. Suitable conditions would have occurred in water, perhaps in the soil (associated with clay particles) or areas where foam and bubbles arise but not, it would seem, on the surface of the seas because of ultraviolet radiation. A recent phylogenetic analysis seeking to establish the composition of RNA in phylogenetic trees favours this scenario, rather than hydrothermal vents, as it suggests that the ratios of certain components in the common ancestor to all life were not those characteristic of organisms in very hot environments.

Banded iron formations and the rise in oxygen

Banded iron formations occur in many parts of the world and constitute the major reserves of iron ore. They consist of alternate layers of red iron oxide and a greyish chert; they were all laid down over about 1750 million years prior to 2000 Mya—a process that William Schopf of the University of California has called the 'rusting of the earth'. The early oceans would have contained much dissolved iron in the unoxidized ferrous state, derived from volcanic activity. As this iron was oxidized the particles of the oxides would have descended on to the sea floor forming the distinctive dull red band. This is usually only a few millimetres thick. Together with the alternating band of chert this probably reflects a strong seasonal shift in the chemical processes oxidizing the iron. Three mechanisms have been proposed for this shift, two of which would result in the iron oxides being laid down in the summer season.

The inorganic mechanism depends on ultraviolet radiation (higher in the summer), causing the ferrous iron to combine with water to form a hydroxide that would be converted to an oxide when the rocks were subsequently heated deep in the crust of the earth. The other two mechanisms depend directly or indirectly on living organisms. One involves certain bacteria, modern examples of which have recently been discovered. These are chemoautotrophs, which use photosynthesis to break down iron carbonate and precipitate iron hydroxide. The other mechanism would involve the straightforward oxidation of the ferrous iron by free oxygen dissolved in the water. This oxygen would have been produced by photosynthesizing cyanobacteria. Fossil evidence reveals that cyanobacteria were active at this time, so the third mechanism must have operated; all three may have played some role, as the amount of oxygen in these deposits is vast, about twenty times that in the atmosphere today.

This process of rusting would have kept the level of oxygen in the atmosphere and water very low, probably with a strong seasonal fluctuation. Bacteria existing at this time are likely to have had to switch between aerobic and anaerobic respiration. Today many species have this ability; one is *Oscillatoria limnetica* and William Schopf has identified very similar species in the Pilbara cherts, some 3500 million years old.

Once this great quantity of iron had been swept from the sea by its oxidation and precipitation, the oxygen produced in photosynthesis would start to accumulate in the atmosphere. This time can be dated by the last of the banded iron formations, namely about 2000 Mya.

The significance of oxygen

When oxygen became a significant component of the atmosphere an ozone (O_3) layer would have formed in the stratosphere, some 15–50 km above the earth's surface. This layer absorbs most of that part of the ultraviolet radiation from the sun that is destructive to many organic molecules and so to life. Water also absorbs this radiation—typically harmful levels occur only in the top few metres of clear water; so the ozone layer is most important for those organisms living on land or near the surface of the sea (or lake).

Oxygen permits aerobic respiration in which one molecule of glucose gives 36 molecules of ATP, the molecule from which energy is obtained. Respiration without oxygen (anaerobic) produces just two molecules of ATP. The greater efficiency of aerobic respiration is essential for the mobility of many organisms. It is also important in the structure of ecological communities where it permits longer food chains. These (FIG. 2.8) describe how one organism eats another that feeds on another and so on until the 'basic industry' is reached. This is most frequently an autotroph (self-feeder) that photosynthesizes. It is often thought that a factor influencing the length of a food chain is the flow of energy, with some being lost each time food passes from one organism to another. Therefore the more energy that can be obtained from each molecule of glucose, the longer the chain. Normally in

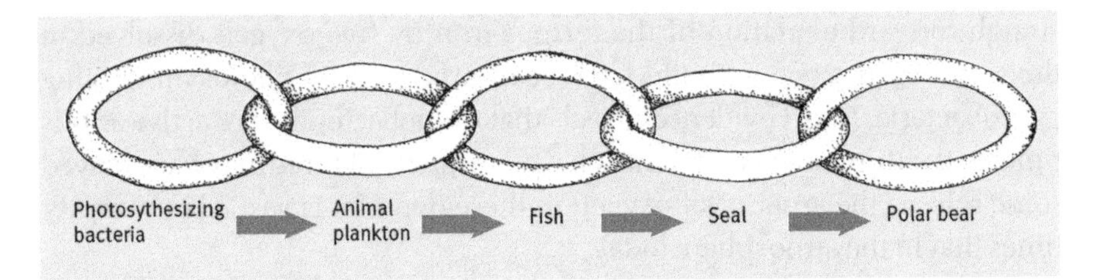

FIG. 2.8 A food chain: the photosynthesizing organisms (autotrophs—self-feeders) at the start of the chain are the primary producers, those in the second link are the primary consumers. All consumers are heterotrophs (feeding on others). In this food chain the polar bear is the top predator

habitats with oxygen there are about five to six links. In anaerobic habitats food chains rarely have more than two links.

As it is so reactive the oxygen ion is highly toxic. Cells of most organisms have come to evolve many special biochemical adaptations to protect them against this damage, though despite this oxygen plays a major role in ageing. However for existing anaerobic organisms, the advent of oxygen would have been a major catastrophe, probably causing the first big extinction. From the viewpoint of the ancient life then existing, the organisms producing oxygen would have been the first wholesale polluters!

CHAPTER 3

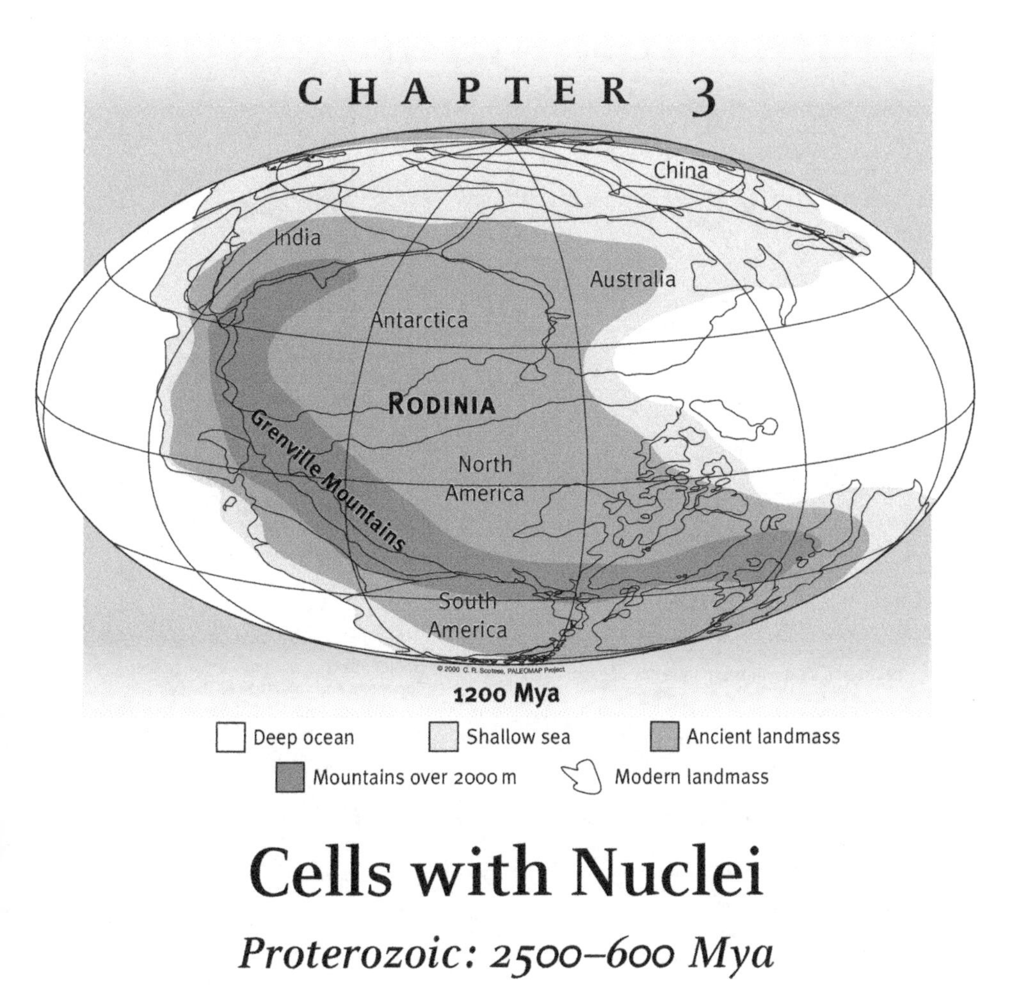

Cells with Nuclei

Proterozoic: 2500–600 Mya

THE development of the nucleus, the feature characteristic of eukaryotic organisms, represents one of the greatest steps in evolution, other than the emergence of life itself. The nucleus can be described as the control centre of the cell. It is bounded by a double membrane and contains nearly all the DNA of the cell, the material that provides the hereditary information. In bacteria (prokaryotes) the usually circular double strand of DNA is attached at one point to the cell membrane; when bacteria reproduce, this strand replicates and so each daughter cell is identical to the parent. In eukaryotes the DNA exists as several pieces held on a number of chromosomes. When the cell divides these form dense structures which, appropriately stained, can be seen under the microscope. In asexual multiplication (cloning) the DNA within each individual chromosome divides to give two

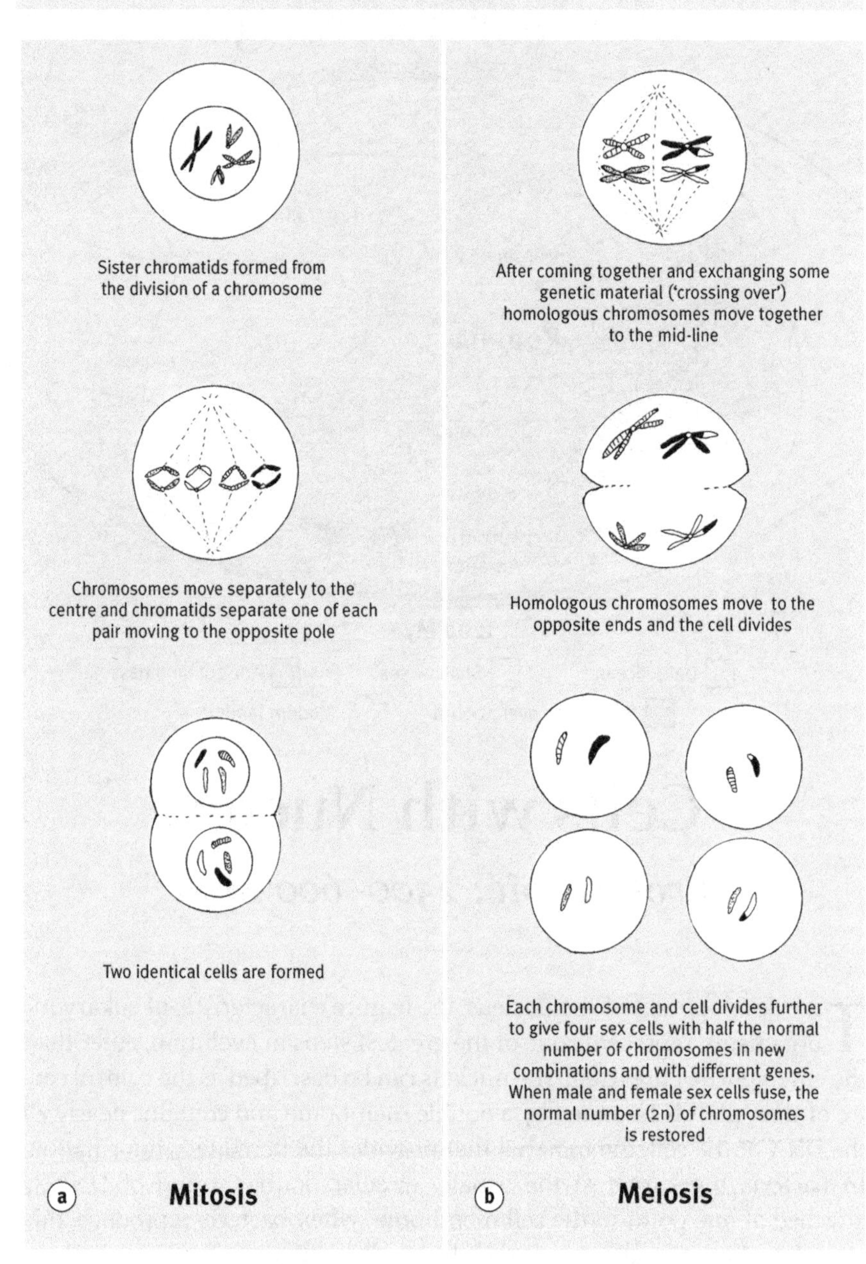

FIG. 3.1 Simplified diagrammatic representation of the two forms of cell division: (a) mitosis and (b) meiosis. The exchange of genetic material during crossing over means that each chromatid is slightly different and ensures that the sex cells have different genetic compositions

identical 'sister' chromatids; these then separate and move to opposite ends of the cell which divides forming two new cells each with the same number of chromosomes as the 'parent' and with identical DNA—a process known as mitosis (FIG.3.1(a)).

In eukaryotes there is another form of cell division (meiosis), which is the basis of sexual reproduction (FIG.3.1(b)). Various rearrangements of the chromosomes occur that result in four sex cells being produced from each parent cell. These sex cells have half the parent's chromosomes, but when they fuse in sexual reproduction the original number (2n) is regained and a fertilized egg (zygote) is formed. This has the potential to develop into a new organism that will have a different DNA composition from either of its parents. Thus, in contrast to asexual reproduction (as in bacteria), sex in plants and animals ensures that every individual of a species will differ slightly and evolution occurs by selection within this almost infinite variety. Meiosis therefore provides the potential for the evolution of very many different types of organism and we might expect therefore that its development must have led to a step change in the rate of evolution.

As well as the nucleus, eukaryotic cells contain other structures called organelles (FIG. 3.2). Some like the cell skeleton (cytoskeleton), cilia and vacuoles are generally considered to have evolved from the components of the original eukaryotic cell ('protoeukaryote') itself, but the mitochondria and chlorophyll-bearing plastids arose from other organisms that became incorporated in the protoeukaryote. When two organisms live together in mutual benefit the relationship is symbiotic and, if one is inside the other, endosymbiotic. The next section describes the serial endosymbiotic theory (SET) of eukaryotic cells.

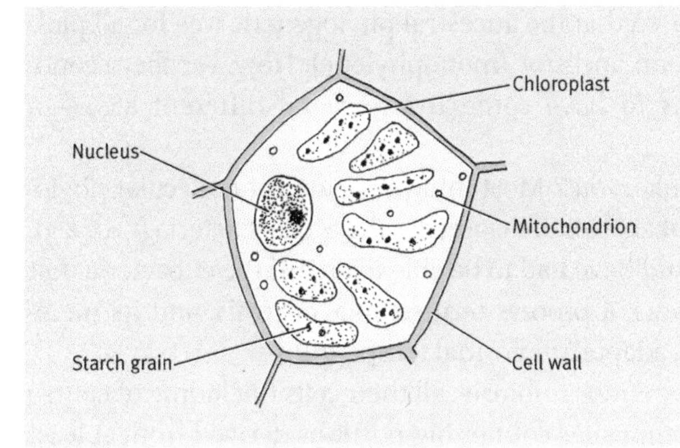

FIG. 3.2 Cross-section of a generalized plant cell showing organelles

Cells—mixtures of organisms

Almost all eukaryotic cells contain mitochrondria, the site of energy production—the 'cell's furnace'. They are surrounded by a double membrane, one derived from the original host (protoeukaryote) and one from the bacterium from which they originated. Most of their DNA has transferred to the host nucleus, but they do retain a small amount. In many cases mitochondrial DNA is inherited almost entirely through the female line and this makes it particularly interesting in some studies of molecular phylogeny (p. 15). All mitochondria probably came from a free-living eubacterium belonging to the large group of α-proteobacteria (FIG. 3.3). It was probably close to rickettsia, which today always live in another organism; one species is the cause of Rocky Mountain spotted fever—a human disease.

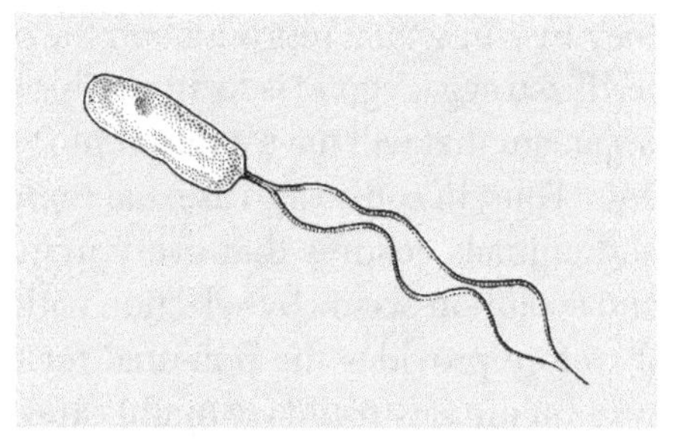

FIG. 3.3 Alpha proteobacterium, *Rhizobium* (after Tudge, 2002)

The green colour of plants, seaweeds and certain single-celled organisms is due to the chlorophyll contained in plastids. These plastids are derived from cyanobacteria, which, like the mitochondria, have retained only part of their DNA. Although often surrounded by two membranes, in many green organisms there are four; here three organisms were involved—the original plastid (cyanobacterium), a unicellular organism (alga) that contained this plastid and a third organism that now contains them both—the chlorophyll is 'second-hand'. The original plastid seems to have arisen from just one species of cyanobacteria, so that the ancestral phylogenetic tree for all plastids has the same common ancestor (monophyletic). However the second-hand chlorophyll seems to have come from several different algae—a polyphyletic origin.

What was the protoeukaryote? Most interpretations of molecular phylogeny indicate that the eukaryotes are closest to the archaebacteria (FIG. 2.5). The protoeukaryote would have had to be able to engulf the eubacteria that became its endosymbionts, a process termed phagocytosis, and its membranes would have been adapted to normal temperatures.

So eukaryotes arose by endosymbiosis: all their cells (including of course those of humans) are composites containing portions derived from at least two, and in green plants, three bacteria. Certain single-celled organisms are exceptions to this as they lack mitochondria. It seems that rather than never having possessed them, in which case they would have represented a stage on the evolutionary pathway to other eukaryotes, they have actually lost them. Their existence shows that a nucleated cell can thrive without mitochondria.

The first eukaryotes were single-celled; collectively such organisms are known as protists. In general those with chlorophyll are termed algae and those without chlorophyll protozoans.

Fossils—a diversity of protists?

The oldest fossils claimed to be those of a protist were found in 1992 in rocks dated as 2100 million years old. They are known as *Grypania* and consist of coiled tubes. These and other filamentous fossils are fairly frequent in more recent rocks (around 1200 million years old), but as little detail can be made out, other than some transverse markings, considerable uncertainty remains as to whether they are eukaryotic algae or complex cyanobacteria.

From about 1700 Mya numerous, generally spherical fossils occur. The earliest ones are simple but those from younger deposits often have elaborate cell walls—sculptured or with spines (FIG. 3.4). They are usually interpreted as the cysts of primitive algae. However, unlike some of the bacterial

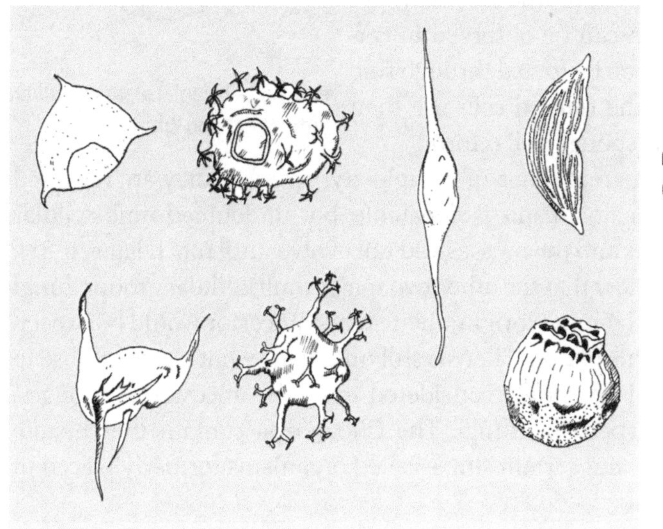

FIG. 3.4 Acritarchs (after Lipps 1993)

microfossils they are not easily placed in modern groups and, reflecting this uncertainty as to their real nature, they are termed acritarchs. The oldest examples are small, a fraction of a millimetre in diameter, but after about 1100 Mya larger forms are found. One, *Chuaria*, can be as much as a centimetre across and is the largest organism found in the fossil record up to this date; it is considerably larger than most of today's unicellular algae. Although acritarchs, especially the larger forms, become rarer in the fossil record after about 900 Mya, recent discoveries in India and China reveal that even *Chuaria* occurred until 570 Mya. A fossil which may be that of a red seaweed (bangiomorphan alga) has been found in the Hunting Formation, Somerset Il., Canada (750–1250 Mya). This apparent diversity of protists during the period 1100–900 Mya probably reflects the evolution of meiosis, that is, of sex.

Many-celled organisms

When they divide certain protists remain together forming a colony (FIG. 3.5); nevertheless the members are separate individuals capable of leading independent lives. In contrast all eukaryotic organisms, other than protists, consist of many interdependent cells; they are multicellular organisms. This represents the next major evolutionary step, an early stage of which can be observed in certain sponges. They can be forced through fine silk muslin, but if the filtered cells are then left in sea water the sponge will reform.

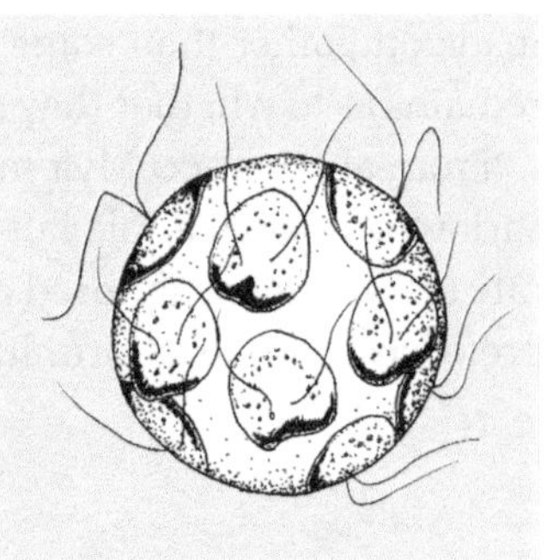

FIG. 3.5 A colonial alga, *Volvulina*. (after Ward and Whipple, 1959)

Seaweeds show a great range of complexity; whether they are regarded as true multicellular organisms is debatable, but undoubted multicellular plants, the liverworts and the mosses, did not evolve until much later (p. 81). Molecular studies show that the other two major multicellular groups, fungi and animals, have a common origin; their shared ancestor would be expected to have lived in this period. There is still uncertainty but a protist close to today's Choanozoa is generally considered to be the ancestor of sponges, other animals and, perhaps, fungi. The Choanozoa contain the choanoflagellates (FIG. 3.6) and certain single-celled organisms formerly placed in the fungi.

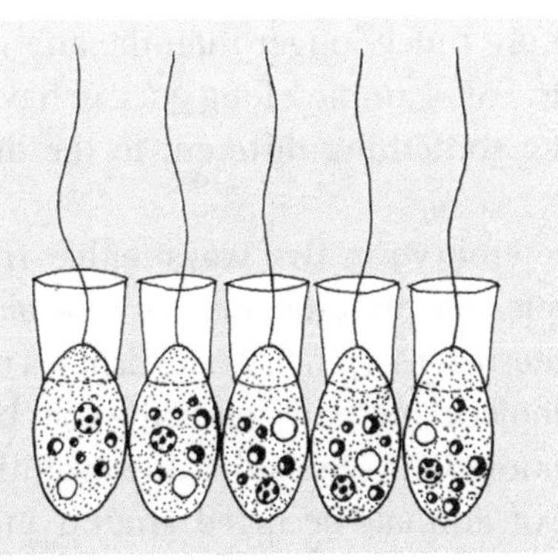

FIG. 3.6 A linear colony of choanoflagellate, *Desmarella* (after Ward and Whipple, 1959)

Animals then went through a number of important evolutionary steps and examples of the different stages still live today (FIG. 3.7). The simplest, but perhaps not the most primitive, animals have their cells in two layers; but the cells in most animals are derived from three layers. The next evolutionary step was the development of a cavity (the coelom) within this middle layer. This cavity is important in providing a space in which organs of, for example, the digestive and circulatory systems can function. Without a cavity they would be squeezed by the locomotory muscles of the body wall, whereas with a cavity they could evolve independent movement and

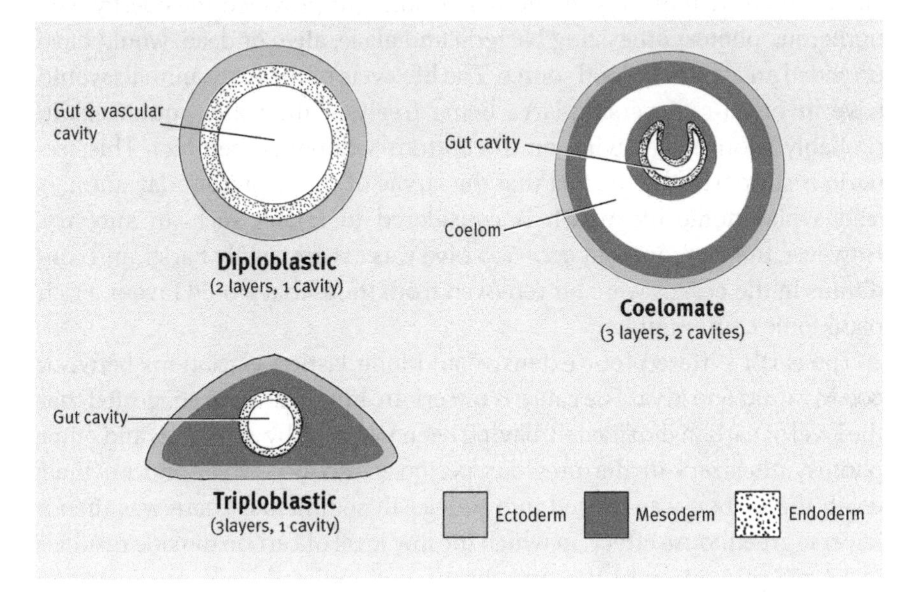

FIG. 3.7 Three major levels of complexity in animals' body plans.

become coiled or folded and therefore much longer than the animal, thus permitting more complex functions. For example a long gut can have many different regions, each contributing something different to the digestive process.

The body cavity develops in the embryo in two ways: either from the front end (Protostomes, for example, earthworms, insects, crustacea, and molluscs) or from the rear end (Deuterostomes, such as starfish, sea urchins, fish, birds and mammals). Recent molecular studies place the split between these two groups at 600 Mya or more. So it seems that the evolution and early fundamental diversification of animals occurred around 1100–600 Mya, although as yet, the fossil record tells us nothing. Such a period of hidden evolution preceding an apparent explosion of new species has been termed a 'fuse'. In this case the so-called explosive appearance in the fossil record of many apparently new types of animal occurs in the Cambrian period (from 545 Mya) (see Chapter 5). But what was the environment in which this fuse burnt, 55 million years and more before the explosion?

Where did animals evolve?

The traditional view is that animals evolved in the seas and oceans in an environment rich in oxygen and where sunlight provided the energy. The numerous photosynthesizing bacteria and algae, alive or dead, would have provided an abundant food source. The life cycle of the early animals would have involved a dispersing larva living freely in the water and an adult, probably sponge-like, living on the bottom sediments (benthic). This scenario is supported by the fact that the larvae of many present day animals lead a planktonic life which is considered to reflect such an ancestry. However, the environment 900–600 Mya was exceptionally harsh, and conditions in the oceans were far removed from those that would favour a rich planktonic community.

The earth suffered four extensive and long lasting glaciations between 900 Mya and 600 Mya. The cause is uncertain, but it has been suggested that the level of carbon dioxide fell, having been taken up by acritarchs and other photosynthesizers in the previous period (1100–900 Mya) and on their death the carbon was carried into the ocean sediments. There was then a 'reverse greenhouse effect', in which the low level of carbon dioxide resulted in a large proportion of the sun's heat being radiated back into space. As snow and ice formed, this increased the reflection back into space, thus creating a feedback. The situation is well described by the term 'snowball earth', for the ice at sea level extended into tropical latitudes. The ratios of carbon isotopes (p. 14) in sediments show that the biological activity in the surface waters of the oceans fell to very low levels. Throughout the period volcanoes were discharging carbon dioxide and other 'greenhouse gases' into the atmosphere. When their concentration reached a critical level, many times the modern level, it is believed there was a sudden reversal in these climatic processes, causing a relatively rapid rise in temperature.

Hydrothermal vents are regions in the ocean floor where superheated mineral-laden water shoots up from deep in the earth's crust. They would have provided a warm habitat for life in 'snowball earth'. Here the available energy is not in the form of light, but of heat, while sulphur and hydrogen, not oxygen, are used in respiration. The most spectacular vents are the huge rock chimneys arising from mid-ocean ridges. Water issues from these 'black smokers' at temperatures as high as 400°C, but the exterior and adjacent regions are cooler. Species of bacteria living around these smokers can withstand, in fact require, high temperatures. *Pyrolobus fumarius*, for example, can survive at 113°C and cannot reproduce at temperatures below 80°C. Living on this rich supply of chemicals and minerals, these chemoautotrophic archaebacteria occur in great numbers: for example, a hundred million *Methanopyrus* can be found in one gram of rock. Reproduction is rapid; at 100°C this bacterium doubles its population every 50 minutes! The same species occurs in the Atlantic as in the Pacific which suggests that these primitive organisms have not changed over hundreds of millions of years.

This dense population of bacteria supports a unique population of often brightly coloured animals. Around four hundred new species have been described since the discovery of hydrothermal vents in 1977. These belong to many different groups, some of them representing types last recorded as fossils some fifty million years ago. Particularly abundant are polychaete worms, mussel-like molluscs and primitive forms of barnacles (FIG. 3.8); the last two produce water currents from which they filter out their microscopic food; often they have apparently symbiotic bacteria living in or on them. Crabs are among the relatively few predators but, as we have already noted, food chains based on oxygen-free environments are short.

There is also another, and much more widespread, anoxic habitat in the sea bed: the black sulphur layer found just below the surface in most marine sediments, whether on beaches or in the ooze in the deep ocean. Besides a

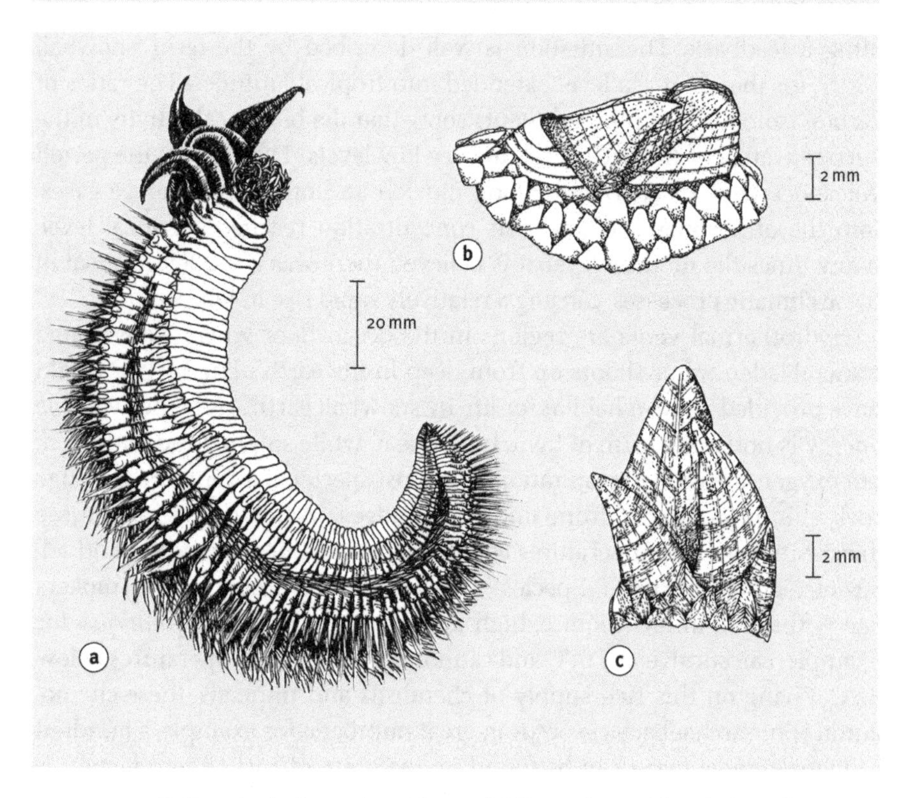

FIG. 3.8 Some hydrothermal vent fauna: (a) 'Pompei worm' (a polychaete), *Alvinella pompejana*, a representative of a previously unknown family; (b) and (c), two primitive barnacles, (b) *Neobrachylepas relica*, 'a living fossil', (c) *Neoverruca brachylepadoformis*, a 'missing-link' between two groups of barnacle (after Desbruyères and Segonzac, 1997)

rich population of chemoautotrophic bacteria, which base their respiration on sulphur, there is a large and varied animal population. Most of these animals lack a cavity (coelom) in their body wall and represent groups that arose towards the base of the evolutionary tree, probably at least before 600 Mya (about the end of the last Precambrian glacial epoch). Many are little known and include certain round worms (Nematoda), jaw worms (Gnathostomulida), flat worms (Platyhelminths) and kinorhynchs (Kinorhyncha) (FIG. 3.9); they are often anaerobic. These are the meiofauna.

Today aerobic animals always have some anaerobic metabolic pathways, but the reverse is not true. It is the conventional view that anaerobic animals have lost the ability to metabolize using oxygen. But if early and large steps

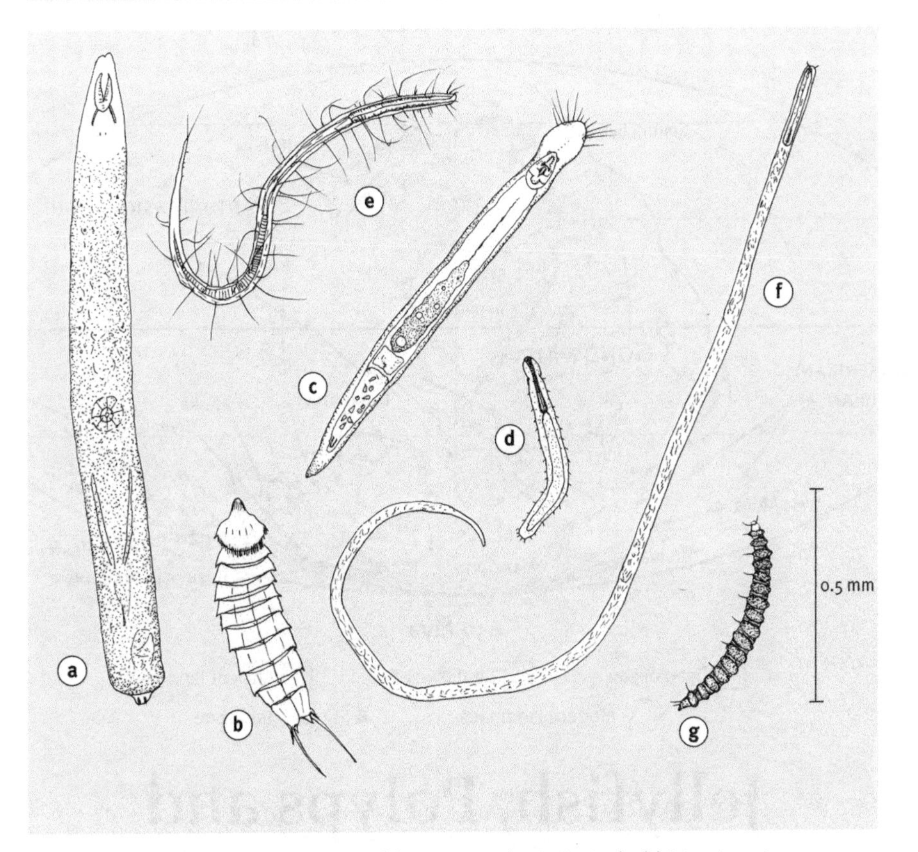

FIG. 3.9 Some meiofauna: (a) flat worm (turbellarian); (b) kinorhynch; (c) jaw worm (gnathostomulid); (d) gastrotrich; (e)–(g), round worms (nematodes) (modified from Platt 1981)

in animal evolution took place, during the snowball earth phase, in anaerobic environments (hydrothermal vents and/or these black sulphur layers) then an anaerobic condition may in fact be the primitive one. Perhaps there was a continuous thread from the very dawn of life in one or other or both of these sulphur-rich habitats on the ocean floor (pp. 10–20).

CHAPTER 4

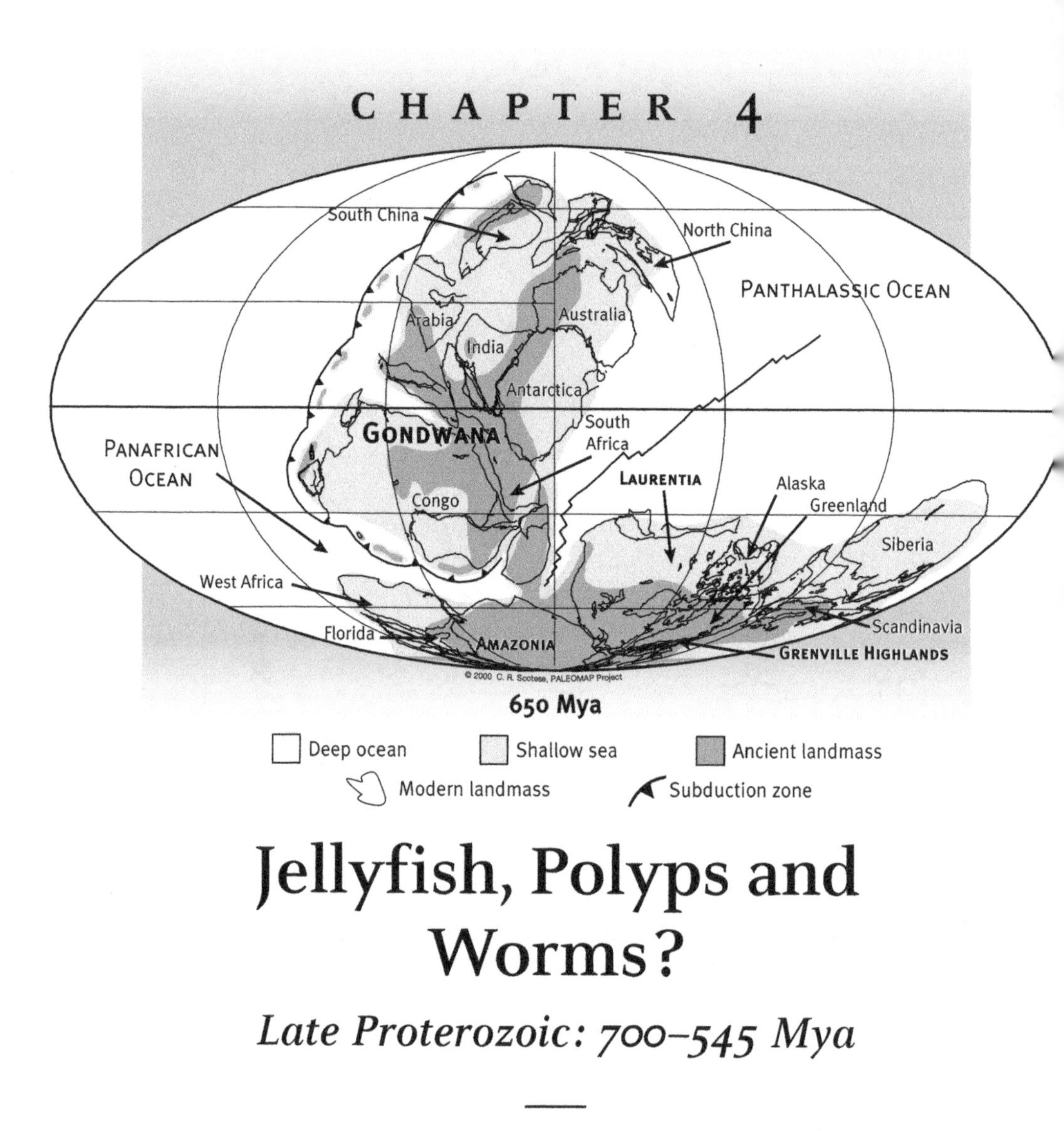

Jellyfish, Polyps and Worms?

Late Proterozoic: 700–545 Mya

The continents of the earth move. The crust (lithosphere) consists today of seven large tectonic plates and a number of smaller ones. Several of them have continents attached to their 'backs', and shuffle across the globe, moved by the deeper convection currents of the liquid rock in the middle layer, the mantle. When the plates move away from each other in the deep ocean, liquid rock (magma) flows upwards and solidifies to form ridges. In some other places the plates may scrape past each other, causing fault lines (such as the San Andreas Fault). When they collide directly, one passes below the other, a process known as subduction. If this happens where the plates are 'carrying' continents then these buckle and rise to form mountains (for example, the Himalayas). Most volcanic activity occurs around the edges of the plates, but some submarine mountains and volcanic islands arise towards the middle of plates. These are due to hot spots, which are generated in particular stationary sites in the mantle. As the plates move across a hot spot a chain of volcanoes are successively 'punched ' through it; only one end of the chain is active—where the plate is currently over the hot spot. The Hawaiian Islands and the older line of submarine mountains, known as the Emperor Seamounts (see map p. 176), that stretch up to the Kamchatka Peninsula, are a spectacular example of the effects of a hot spot and show how the Pacific plate has moved over the past 75 million years.

This 'continental drift' has only been widely accepted and understood in the last fifty years. Many lines of evidence favour the idea of plate tectonics, but the key breakthrough was made through studies of rock magnetism. When a rock solidifies under the ocean ridge the iron minerals within it take on a magnetic orientation and strength that reflects the geographical relationship it then has to the magnetic pole. For unexplained reasons the earth's magnetic field reverses at irregular intervals of about a hundred thousand to a million years; that is the compass needle reverses its direction and for a period our 'north' points to the south pole. As the plates have moved apart these shifts have generated a series of alternating magnetic strips on the ocean floor, disposed symmetrically on each side of the ocean ridge. By dating rock samples from the magnetic strips we can find the frequency of magnetic reversals over the last 200 million years. This tells us how the oceans have spread over this period, but most of the past of the ocean floor that is older than this has been subducted into the interior of the earth. Now that it is known that the tectonic plates move it is possible to determine the position of the continents through geological time from a knowledge of the exact direction of magnetization of iron-bearing volcanic and sedimentary rocks which form part of the continental crust and from comparisons of continents that were once joined and have now drifted apart. Additional information used in interpretation is gained from the features of ancient mountain ranges, and from the nature of sedimentary rocks, laid down under the sea, and from the fossils they bear. We can gain an idea of the depth of the sea in which sedimentary rocks were deposited and sometimes tell whether this was in a tropical or polar region. In this way maps have been built up of the geography of the world in the past.

Twice in the last 1200 million years the tectonic plates have moved so that their overlying burdens of continents have come together to form supercontinents. The first formed has been named Rodinia (see the map at the start of chapter 3) and the second is known as Pangaea. Further movements of the plates have caused these supercontinents to split up to form once more several separate continents. The various stages of this process have very different effects on the global environment. When the continents collide the crust comes under great compressive forces, and is pushed up to form mountains. At the same time the continents themselve become shorter, increasing the area of the oceans, causing the sea level to fall. When they move apart the crust comes under tension, being pulled out, in principle reducing the ocean area and causing the sea to rise. The sea level at any one time depends on two factors. Firstly, by the extent to which the crust is pushed above the ocean—like someone sitting-up in the bath! And second, by the sizes of continental ice sheets. Various combinations of these processes have led to a continuous change in sea level throughout the earth's history (FIG. 4.1).

Because climate on land is so influenced by the closeness of the sea and the nature, warm or cold, of the offshore ocean currents, the size and shape of the continents has a profound influence on the environment. For example, when the supercontinent was formed much of the land would have

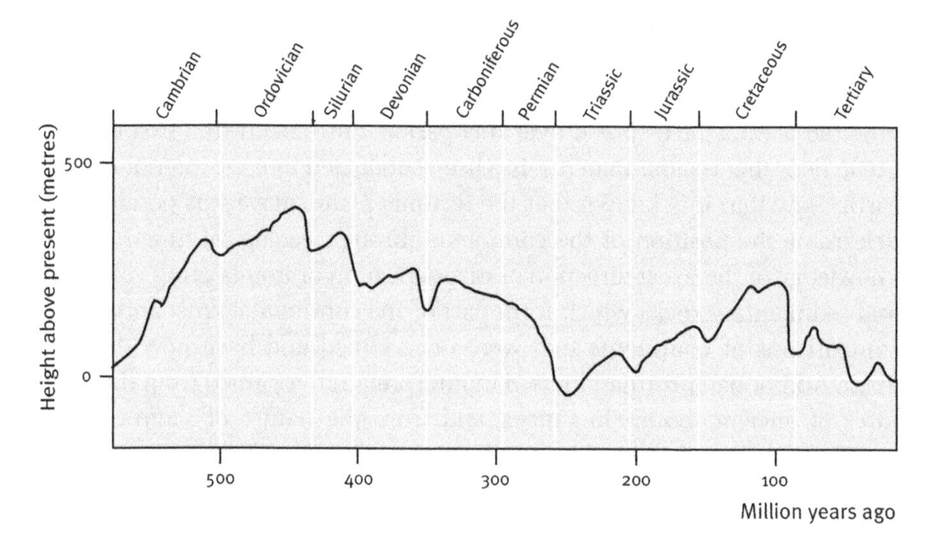

FIG. 4.1 Global sea levels (largely based on Elliott, 2000)

been far from the sea and would therefore experience extreme continental weather (hot and dry in summer and cold in winter). As the maps at the start of each chapter show, the latitude of a particular piece of land is always changing as it moves around on the plates. Some areas now in the tropics have in the past sat over the poles. Clearly geographical position will have a strong influence on the local climate, but the overall global climate is very dependent on the extent to which large land masses are in the polar regions. As we saw in the previous chapter, changes in the chemical composition of the atmosphere, especially the proportion of carbon dioxide, also have a profound influence on climate.

The supercontinents had two major components: Gondwana and Laurasia. Australia, India, Africa, South America and Antarctica were in Gondwana; and North America, Greenland, Europe and much of Asia in Laurasia.

From about 700 Mya the early supercontinent is believed to have been splitting up. Within the next 100 million years—as the era of great glaciations responsible for 'snowball earth' drew to a close—there were extreme fluctuations in sea level. It seems that, after the initial split, Gondwana—which then lay towards the north—reassembled. From about 590 Mya there was probably a period of relatively mild climate with many seas between the fragmented land mass. Did this provide a new opportunity for life?

The Ediacaran world

Examining old silver-lead mines in the Ediacaran Hills in South Australia in 1946, Reginald C. Sprigg discovered numerous fossils that resembled impressions of jellyfish; he recognized that these were perhaps the oldest known animal fossils, but it was not until further collections were made there some ten years later that their full significance was realized. Here was a fauna very different from most known animals and it came from a late period of the Precambrian that had previously been thought to be devoid of fossils of a size that could be seen with the naked eye. In the last fifty years similar fossils have been found in rocks of Ediacarian age (590–545 Mya) in several parts of the world, including Namibia, China, Russia, northern Europe and Newfoundland. In these sites, which lay around the edge of the Panthalassic Ocean, many of the organisms found were the same. The 'Ediacaran fauna' was widely distributed. Its components can be loosely grouped into three different designs (FIG.4.2).

FIG. 4.2 Some Ediacaran animals (modified from Glaessner, 1984)

A number of them (for example *Ediacaria, Eoporpita*) resemble jellyfish and related animals, being circular with a series of radial divisions. They are termed radially symmetrical—they have no left or right side; as with a wheel any line through the middle will divide them in half. They were probably mostly free floating, but some may have rested on the sea bed; both, it is believed, would have fed by trapping particles (protists, small animals, etc.) on their filamentous tentacles. Some are large, the size of a dinner plate.

Charniodiscus and other frond-like forms may have been even larger, reaching a metre or more in length. They bear some resemblance to modern sea pens; they probably lived attached to rocks on the sea bed, but some may have been free-floating. It is likely that they also filtered out their food from the passing sea.

The most varied group can very loosely be described as 'worm-like'. They are bilaterally symmetrical: that is they may be considered to have right- and left-hand sides that define the mid-line. Three widespread types are *Spriggina, Dickinsonia* and *Kimberella*. The first named has a distinct head and a body that has divisions like the segments of an earthworm. *Dickinsonia* has a quilted appearance and some resemblance to a living ectoparasitic polychaete worm. Well-preserved specimens of *Kimberella* recently described from rocks dated 555 Mya suggests that it is reasonable to speculate that it had three layers of body cells (see FIG. 3.7).

In addition to the fossils of actual animals that exist from this period, there are many fossilized tracks made by animals burrowing within the mud. There are also tubes in which creatures like a coral polyp could have lived; one frequently found is called *Cloudina*. Acritarchs and recognizable algal remains are often found in fine-grained sediments. So the Ediacaran world had many inhabitants. How did they live and how are they related to organisms that evolved later?

There is a great deal of controversy about the answers to these questions, a reflection no doubt of our relatively poor knowledge. One can be confident that further work and discoveries will clarify the picture. The fossilization of Ediacaran animals is something of a conundrum, for none had a skeleton. It may be that rather special physical conditions existed at this time and the fact that these animals appear so infrequently later on in the fossil record was because conditions for fossilization had changed, not because they were not there. Another suggestion made to explain the apparent uniqueness of their fossilization is that the saprophytic way of life (that is consuming dead material) had not evolved; but this seems improbable in the light of the already existing diversity of bacteria and protists. Another line of argument is that without a skeleton they would have had little defence against predators and this feature signalled their death knell when predators evolved after the development of claws and jaws (see Chapter 5). This has led to the view that the Ediacaran world was a gentle one with large jelly-like animals living on small particles or obtaining their nutrition from symbiotic alga—the 'Ediacaran Garden'. Against this rather idealized view must be set the facts that modern jellyfish and their relatives have stinging cells (nematocysts) that can be lethal to organisms larger than themselves. Furthermore holes, that look as if they were made by a boring predator, have been found in the sides of *Cloudina* tubes.

However, some would argue that the Ediacaran fauna was a quite distinct form of life and unrelated to those that followed: it was an evolutionary dead-end. The leading protagonist of this theory is the German palaeontologist Adolph Seilacher. He views the apparent similarities with other known animals as misleading and considers that they left no modern descendants. He has proposed that the Ediacaran fauna should comprise a new group called the Vendobionts (after the last geological interval in the Precambrian, the Vendian, in which these fossils occur). He believes that the body walls of the vendozoans were made of tougher material than that of sea pens, jellyfish and many worms; this is why they were fossilized. Rather than standing up from the floor of the sea, *Charniodiscus* and similar forms would have often lain on the surface mud or fine sand, obtaining their nutrition from symbiotic photosynthesizing algae. In fact, the latter feature is not incompatible with their being related to modern animals, rather than being an aberrant dead-end; such relationships between animals and algae are widespread today, for example on coral reefs, where both the polyps, that form the reefs, and the giant clams live symbiotically with certain algae.

As will be seen in the following chapter, many fossils from the next geological period initially appeared to represent animals totally distinct from subsequent faunas but, as more knowledge has been gained, their true identity has been revealed and they can be related to the accepted evolutionary tree of animals. It seems that the same may apply to the Ediacaran fauna. Some interpretations place the worm-like forms as primitive arthropods, others suggest they may be early annelids. A conclusion from studies of molecular phylogeny is that animals that were ancestral to both groups, and so having characters of both, would have lived during this period. If these are they and not some unique dead-end, one would expect some to survive into the Cambrian. Indeed this seems to have occurred as fossils resembling *Charniodiscus* have been found by Simon Conway Morris of Cambridge University in the Burgess Shale of that period.

At around 545 Mya the geological record shows abrupt changes in the material deposited on the sea floor that clearly reflects changes in the environment. Small shell-like fossils appear in large numbers. A new era had begun.

CHAPTER 5

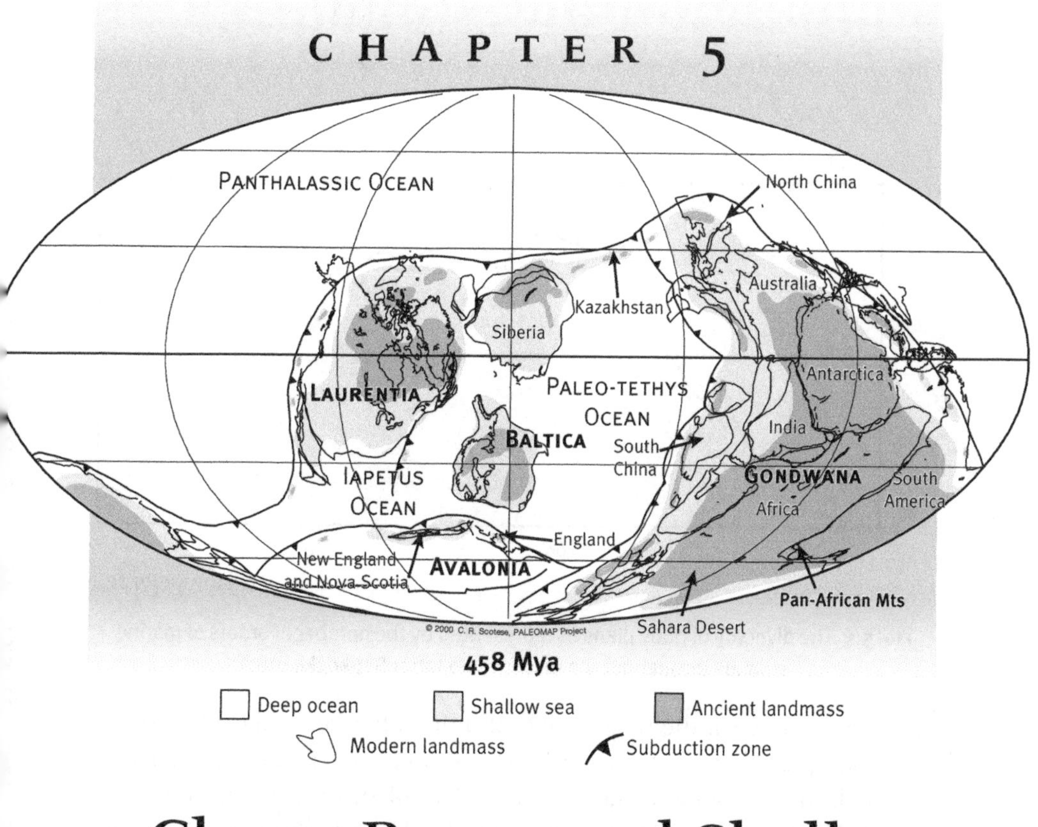

Claws, Rasps and Shells

Cambrian–Ordovician
545–438 Mya

THIS is the one period in the whole of the history of life when the number of different animal forms—their diversity—appears to have increased most dramatically. This is shown for marine animals in FIG. 5.1, where the measure used is the number of orders. An 'order' contains species of animal of similar design (p. 4). Different species of crab are in one order (Decapoda), of woodlice in another (Isopoda); both belong to a bigger category, the class, which in this case is known as Crustacea. This sudden burst of evolution has been termed the 'Cambrian explosion'. However as has already been mentioned, molecular studies point to many fundamental

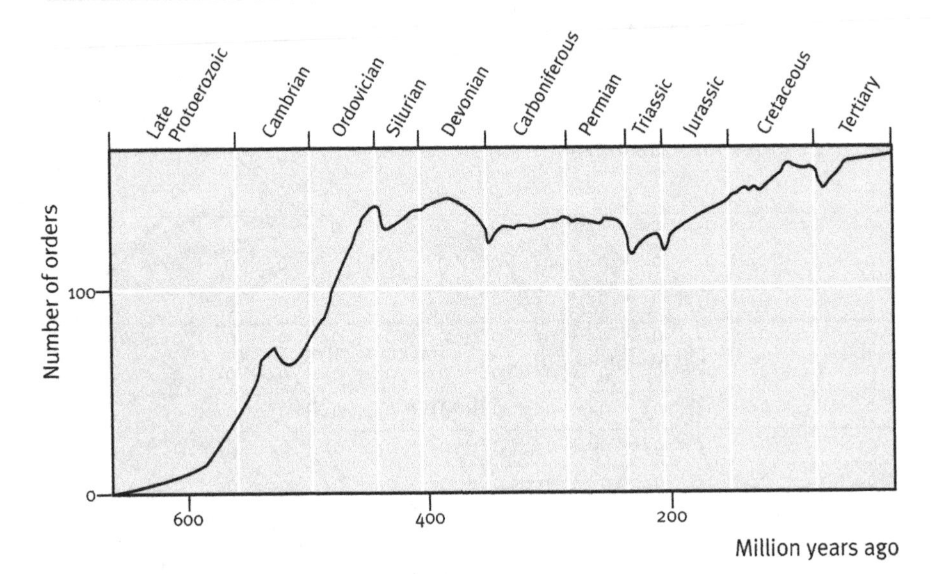

FIG. 5.1 The diversity of body plans as represented by the number of orders of marine animals throughout the fossil record (after Sepkoski, 1981)

differences between the major types of animal having evolved before the Cambrian period. Also, if certain interpretations of Ediacaran animals are accepted, there is some evidence, supportive of this view, from the fossil record. The previous (Precambrian) period has been described as a 'fuse', the time when the potential for an explosion was established.

But what ensured that the fuse would ignite the explosion? The immediate cause seems to be that animals developed skeletons; this step facilitated the evolution of claws and biting or grinding mouthparts (like those of a modern crab) or drilling or rasping mouthparts (like those of modern snails): important equipment for the predatory way of life—killing and consuming other organisms. The holes in *Cloudina*, referred to in Chapter 4 (p. 41) reveal that drilling had already developed in the fuse period. Evolution would lead to the acquisition of defences, either shells to provide protective armour, or various muscular systems to allow animals to move with speed. Some form of skeleton is required for speedy movement, for if muscles are to be effective they must be anchored to something firm to contract and pull against. As so much of our understanding of past life depends on fossils, we should perhaps note, that whereas strong bones and robust shells are relatively likely to fossilize, other skeletons, (like the shrimp's) which are perfectly effective for the living animal, have a lower chance of preservation.

But what triggered the development of skeletons? During the Cambrian period some animals, protists (single-celled organisms) and seaweeds started to incorporate minerals in their body tissues, a process known as biomineralization. It seems most likely that there was some change in the environment, perhaps a rise in the oxygen level, that facilitated the physiology of biomineralization and led to the formation of mineralized skeletons.

It can also be argued that the increase in the richness of the fossil record at this time is due to conditions that favoured fossilization. Naturally skeletons have a greater probability of surviving to become fossils than the soft parts of organisms, so this argument also depends on the advent of biomineralization.

Taking in minerals—biomineralization

Two types of process enable an organism to incorporate minerals. One is termed 'biologically induced' and occurs when the condition of a cell causes a mineral to precipitate on it. Therefore, the type of biomineral formed in this way reflects the surrounding environmental conditions. The other is known as 'biologically controlled' and as the name implies depends on special conditions in the cells and the mineral crystals form within a network or matrix laid down by the cell. There is precise biological control in this process which shows a measure of independence from the environment; for example, a relatively rare chemical may be differentially taken in from the environment. Temperature often affects the extent of mineralization; many marine organisms lay down more calcium minerals in warm waters than in cold waters. The bones of arctic fish are therefore less calcified than those of fish from temperate or tropical seas.

A variety of chemicals are incorporated in organisms by these processes. The one most frequently taken up is calcium carbonate, which forms minerals like calcite and aragonite that are widespread in skeletal structures. Fewer animals use calcium phosphate, but it is the major component of most vertebrate skeletons. Silica occurs in many protists, certain invertebrates (like sponges), and some plants (like bracken—making its broken stems sharp enough to cut ones hand). Oxides of iron, particularly in the form of magnetite and goethite, often have special functions, ranging from responding to magnetic fields to providing a particularly hard tip for a mouthpart like the molluscan radula (used as a rasp or as a drill) (see FIG. 5.2).

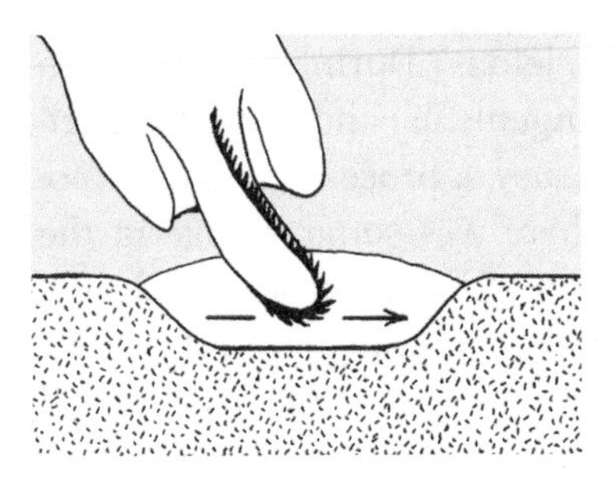

FIG. 5.2 The rasp (radula) of a molluscan drilling into the shell of another animal (after Younge and Thompson, 1976)

The dawn of a new era

The fossil record in rocks of this age (Cambrian 545–505 Mya) is very different from that of older formations. The first sign of this change is the presence of many shelly fossils (FIG. 5.3), which are very small, often a millimetre or less. Some of them, like *Aldanella* seem to be minute snails, yet others are very puzzling: they may be only parts of an animal. Calcified cyanobacteria are found for the first time, yet this a group that has existed virtually from the origins of life and in general shows little evolution—the Cambrian species are not very different from those found today. A contrast to this long survival is provided by the sponge-like archaeocyathids (FIG. 5.4). They first appear at the very beginning of the Cambrian but, having become rare by the middle Cambrian, disappear altogether (presumably having become extinct) by the end of the period.

By the middle of the Cambrian (around 530 Mya) a very much richer fauna existed in the seas; the Cambrian explosion had manifested itself in a diversity of body plans (sometimes described as disparity). This diversity of body plans in marine animals continued to increase (FIG. 5.1(a)) throughout

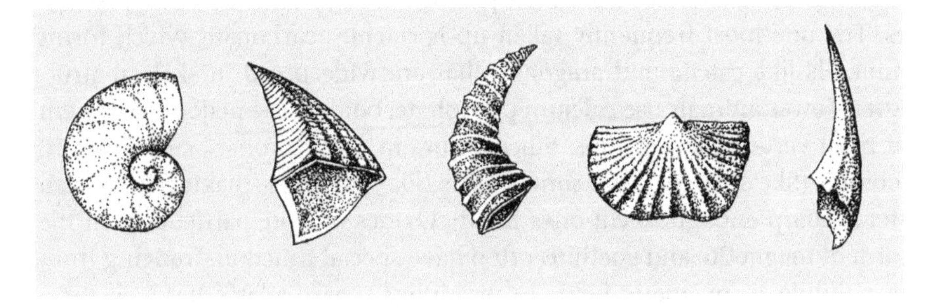

FIG. 5.3 Examples of small shelly fossils from early Cambrian (approximate size range 0.7–2 mm.) (after Clarkson, 1998)

the next geological epoch, the Ordovician (505–438 Mya), and then plateaued out—albeit punctuated by a series of extinction events. However, although the fundamental types of basic body plan did not increase after the Ordovician, the number of families of organism most certainly did (FIG. 5.17).

Much work has been undertaken in seeking to unravel the factors that led to this diversification in the Cambrian. Perhaps the trigger was internal to life. It could have been the effect of a new type of gene: those in HOX clusters. These are genes that, together with their derivatives, play a major role in development, controlling the growth of different structures in relation to their position on the body: eyes on the head, hind legs near the tail. This provided the possibility for whole new suites of animal architecture. However, it seems that these genes evolved before or with the body cavity (termed the coelom cavity, FIG. 3.7), and molecular phylogeny (ancestral relationships based on molecular structure) suggests that this would have been well back in the Precambrian 'fuse' period. For the immediate trigger we come back to the role of environmental factors.

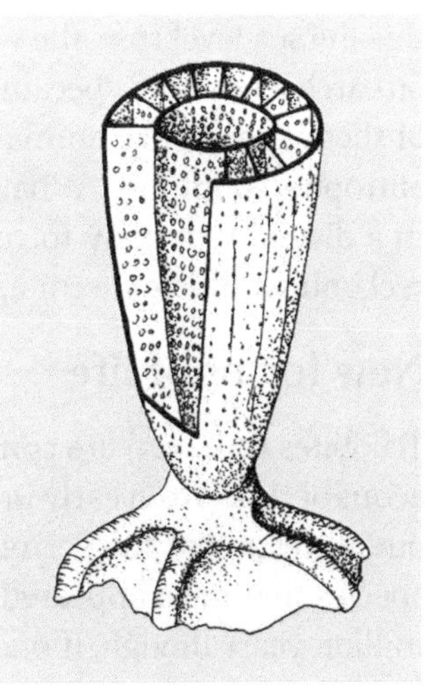

FIG. 5.4 A sponge-like archaeocyathid (approximate height 4 cm) (after Clarkson, 1998)

At the very start of the period the sea level was low but the water was rich in phosphorous. Many of the small shelly fossils from this time have shells made not simply of calcium carbonate (as is the general rule throughout the rest of evolutionary history), but also include phosphorus. In some fossils the calcium–phosphorus complex was the material of the shell of the living organism; in others it is considered that it replaced the calcium carbonate during the process of fossilization. Phosphorus is an important nutrient and waters rich in it become 'eutrophic'—that is, a great population of photosynthesizing plankton develops and when these die and decay, the oxygen is used up so that aerobic organisms cannot live. Today, eutrophic conditions are often the result of pollution by sewage and other wastes rich in phosphorus and nitrogen; the water first becomes green and then the fish die. Perhaps these eutrophic conditions in some localities in the early Cambrian favoured anaerobic animals from the meiofauna (p. 34). After this the sea level rose, the waters were no longer eutrophic, and the calcareous archaeocyathids became widespread. It may be that the near-extinction of these sponge-like animals later in the period was due to a recurrence of eutrophic conditions. Whatever the reasons, from this period life flourished in a diversity of body forms that had never occurred before and, in some reckonings, did not occur again.

New forms of life

Trilobites (FIG. 5.5) are conspicuous members of the fossil fauna for many aeons of time, from early in the Cambrian (545 Mya) until they finally faded out at the end of the Permian (248 Mya). There are thousands of different species that have appeared and disappeared at various times in these 300 million years though, if one is to speak of an 'Age of the trilobites', it is this

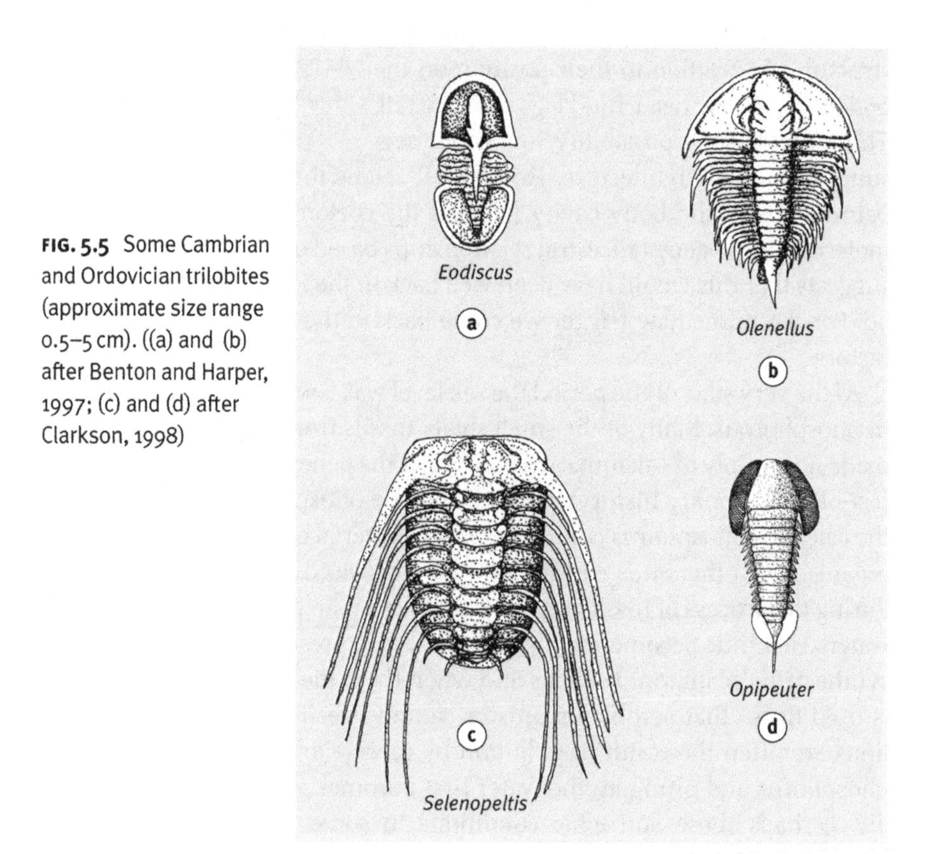

FIG. 5.5 Some Cambrian and Ordovician trilobites (approximate size range 0.5–5 cm). ((a) and (b) after Benton and Harper, 1997; (c) and (d) after Clarkson, 1998)

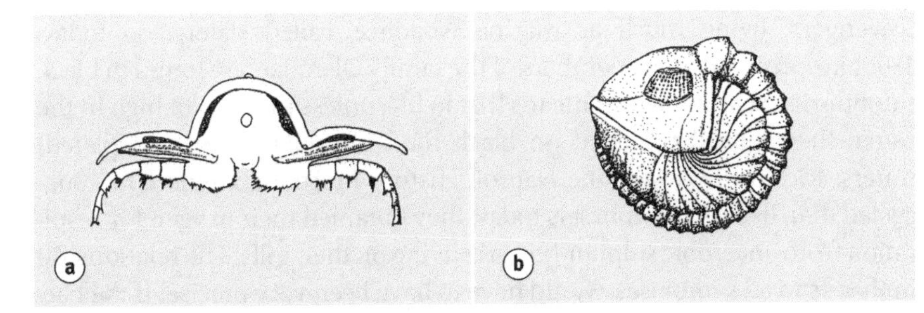

FIG. 5.6 (a) Cross section of a trilobite (after Whittington, 1975); (b) trilobite in 'rolled-up' position (after Clarkson, 1998)

one—the Cambrian and the Ordovician. Trilobites are commonly around five centimetres in length, but range from a millimetre to over half a metre. They belong to the great phylum Arthropoda that contains animals with an exterior skeleton and jointed legs; today members include woodlice (to which trilobites have a superficial resemblance), crabs, prawns, insects and spiders. Trilobites have many special features; as their name suggests the body is divided longitudinally into three lobes, it is also divided into three main regions horizontally; the middle region is further divided into segments. The body bears many appendages: besides antennae on the front and sometimes the cerci on the tail end, there are paired appendages on every segment—these are all of the same basic two-branched structure (FIG. 5.6(a)). The upper part is commonly called the gill, though it is believed to have often served functions in addition to respiration, for instance swimming, when it took a paddle-like form. The lower part of each appendage, the leg, was for walking. The base of the appendages, particularly those on the head, often bear strong spines which almost certainly served as jaws for holding and grinding food. The unique feature of trilobites is the lens of the eye: this is made of the mineral calcite, as is the whole exoskeleton. In all other animals the eye lens is proteinaceous. The eye generally has many facets, often many hundreds. It is possible to look through the fossilized eye using a camera; this has revealed that in most species front and side vision were particularly good, a result achieved because the crystals of calcite are all aligned in a particular direction—an example of the precision of biologically controlled biomineralization.

Most trilobites lived on the sea bed, and the projections of the body may have acted like a snow shoe, spreading their weight over the very soft muddy surface. Although some, like *Olenoides* were predators, many were scavengers, living much as marine woodlice, called slaters, do today. Trilobites, especially the members of the family Olenidae, are found in black sulphur-rich shales—which means that in life, unless they swam high in the water, they must have lived on black mud in anoxic (oxygen-depleted) waters. Richard Fortey of the Natural History Museum, London, has suggested that, like certain mussels today, they obtained their oxygen for respiration from anaerobic sulphur bacteria living on their gills. The relationship in this, as in all symbioses, would have to have been very precise: if the bacteria produced more oxygen than the trilobite could use they would poison themselves, if too little the trilobite would die. This type of relationship, where the waste product of one organism is poison to itself but 'the breath of life' to another is also the basis of bacterial consortia (p. 21).

When crawling along the sea bed a trilobite was well protected by the armour of the thick exoskeleton on its back; but turned over, with its undersurface exposed, it would be very vulnerable. However, fossil trilobites show that, like the modern woodlouse or pill bug, some of them could roll up in a ball (FIG. 5.6(b)). After the Ordovician period those trilobites that could not roll up became rarer.

The hard exoskeleton cannot stretch as the animal grows so, like all other arthropods, trilobites would have moulted, shedding their old skin and expanding quickly while the new skin was still flexible. Their detailed structure would have changed from one moult to the next (FIG. 5.7). The early stages were very small and some may have led a pelagic life as part of the plankton in the upper waters of the oceans; this is the lifestyle of the larvae of many modern arthropods such as crabs and lobsters that as adults are bottom dwelling (benthic). Certainly some trilobites like *Opipeuter* (FIG. 5.5) lived a pelagic life, even as adults.

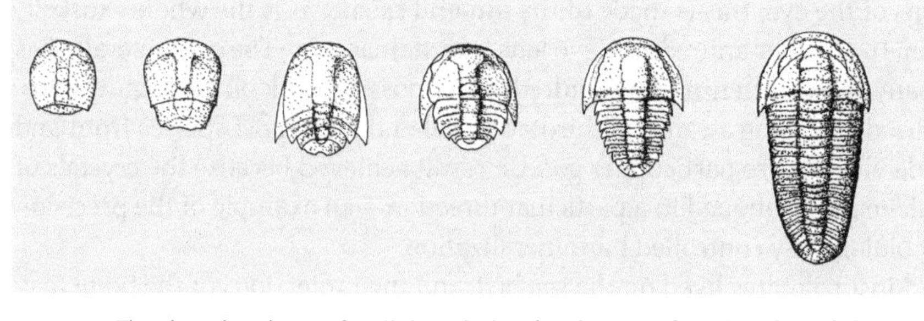

FIG. 5.7 The changing shape of a trilobite during development from larval to adult stage (after Clarkson, 1998)

The sea scorpions (Eurypterida) (FIG. 5.8) are also arthropods and, as with trilobites, the group is now extinct. Appearing in the Ordovician, they were predators and presumably caught their prey with the powerful claws on their front legs. Some specimens were two metres in length and were the largest arthropods ever to exist. The existence of such a large predator suggests that the food chain had extended, perhaps to its approximate maximum of five links (see Chapter 2, FIG. 2.7).

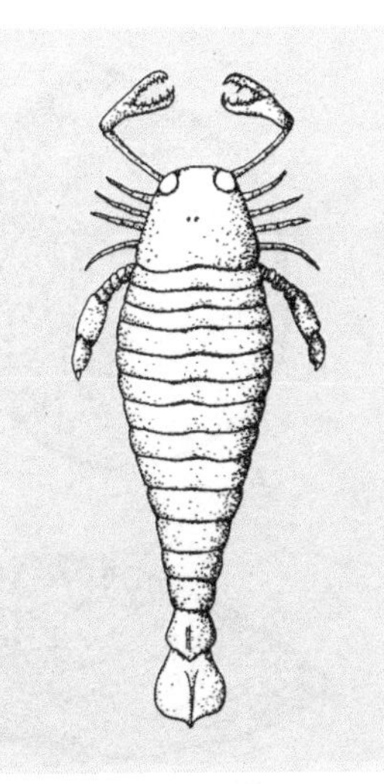

FIG. 5.8 A sea scorpion (*Pterygotus*) (approximate length 2 m) (after Clarkson, 1998)

Two groups of molluscs are well represented as fossils in this period: they are Gastropoda (snail-like forms) and Cephalopoda. The nautilus (see Chapter 8, FIG. 8.9), octopus, cuttlefish and squid are living cephalopods and the group has a strong presence throughout much of the fossil record. Whereas the nautilus and ammonites (FIG. 8.8) have coiled shells, in others, including those found in the late Cambrian and Ordovician, the shells are cone-shaped. Rapid propulsion is achieved in many by a jet of water forced out through the siphon (a muscular tube). All cephalopods are predators, typically tearing their prey with their horny beak. They may have been responsible for causing the bite marks often found on the sides of trilobites; they were almost certainly major predators in the seas, some having shells three metres in length.

Most marine snails (gastropods) feed by grazing on algae; they scrape up the material with their radula, a mouth part like a fine rasp. Some are predatory, drilling holes through the shells of other animals (FIG.5.2); drill holes have been found in the Precambrian *Cloudina* and in some Cambrian lampshells (Brachiopoda). Other groups of mollusc also occur as fossils from the Cambrian. Particularly interesting—because their structure may hint at the origin of molluscans—are the monoplacophorans (FIG. 5.9) that have their body muscles divided as they are in segmented animals like worms (annelids) and Arthropods. This group was thought to be extinct until members were discovered in the deep ocean in the 1950s. There are several examples of living fossils which have been found in the deep ocean, sites that one would expect to be least affected by the sorts of events (widespread volcanoes, glaciations, asteroid collisions) that have caused spasmodic periods of

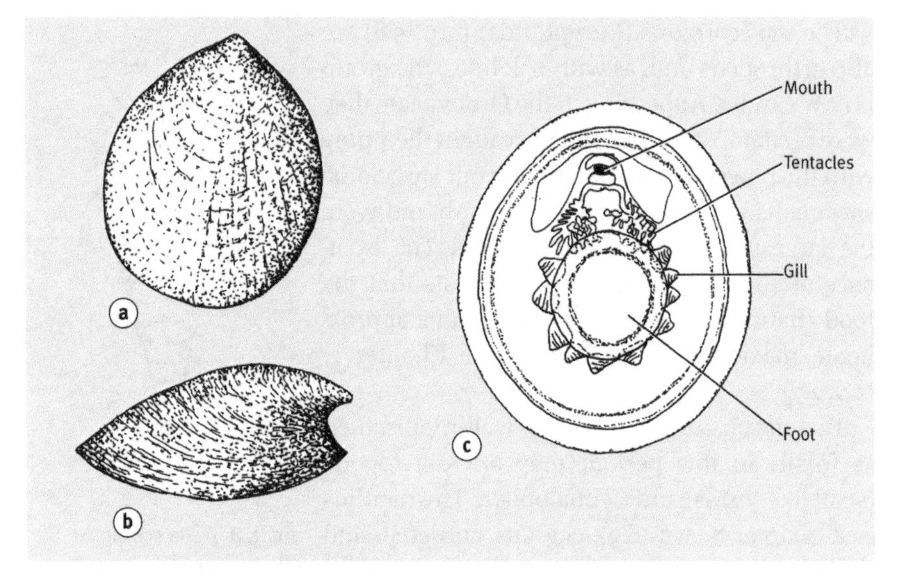

FIG. 5.9 A living monoplacophoran, *Neopilina*: (a) and (b) views of shell, (c) view from underside ((a) and (b) after Clarkson, 1998; (c) after Meglitsch and Schram, 1991)

extinction on the earth. A further group of molluscs, the bivalves (mussels, cockles, scallops, oysters, etc.) became common towards the end of the Ordovician period. They have two shells, in some cases identical, but in others very different in shape. They live in sand or in mud, or attached to rocks, and mostly feed by creating a water current that they then filter for food.

Lampshells, brachiopods (FIG. 5.10), were very abundant in the Cambrian and Ordovician seas. They have two shells and superficially look like bivalve molluscs (which were rare in the Cambrian and early Ordovician); in some cases these shells are hinged, in others held together by muscles. They live attached to a rock or buried in the sand or mud and extract small particulate food (e.g. protists, bacteria) from a current of water that they draw in and over a tentacle-covered organ known as the lophophore, which also serves as a respiratory organ. This structure is quite unlike anything seen in molluscs, which demonstrates that these groups are of fundamentally different design. The superficial resemblance of lampshells to bivalves that live in the same habitats is an example of convergence meaning that two groups of unrelated organisms can often look alike because of similar adaptations which have evolved to allow them follow similar modes of life. Although very abundant at this period of geological history, lampshells

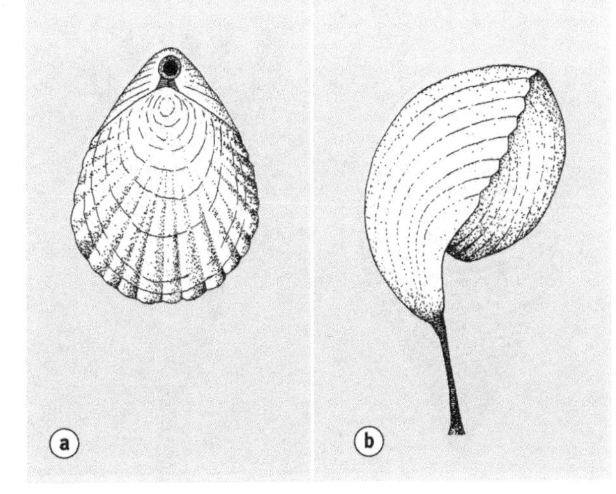

FIG. 5.10 Lampshell (Brachiopoda): (a) lower surface, (b) side view in life showing pedicle by which it attaches itself to some substrate (after Clarkson, 1998)

became rarer thereafter, and only a relative handful of species occur today. One of these, a *Lingula*, is very similar to species that occurred in the Cambrian and so, among the many-celled organisms, this animal holds the record for survival—about 500 million years. So far *Homo* has lasted for at most three million years. *Lingula*'s ability to tolerate both anoxic and brackish conditions has undoubtedly been important for this survival.

Fossils of yet another group are often found, the Echinodermata—literally 'the prickly skinned ones'. Living examples are sea urchins, starfish, sea lilies and sea cucumbers. None of the groups found in rocks from the Cambrian exist today. One, the curious helicoplacoids (FIG. 5.11(a)), seems to have been extinct by the middle of the period. The sea lilies, crinoids (FIG. 5.11(b)) suddenly became very widespread and abundant in the Ordovician and remained a major component of the marine fauna until the end of the Permian, after which they were less abundant and diverse.

Reefs

Marine reefs are particularly interesting for two reasons. Firstly, they provide more 'ecological space—ecospace'. Where there would otherwise be a surface of rock or sand with a column of water above it, the reef provides a complex three-dimensional structure with a great amount of extra surface, with many cracks and small caves. In modern reefs the fore-reef is exposed to the ocean waves, whilst the back reef is sheltered. Thus reefs supply a wide range of habitats and so it is not surprising that their biodiversity is

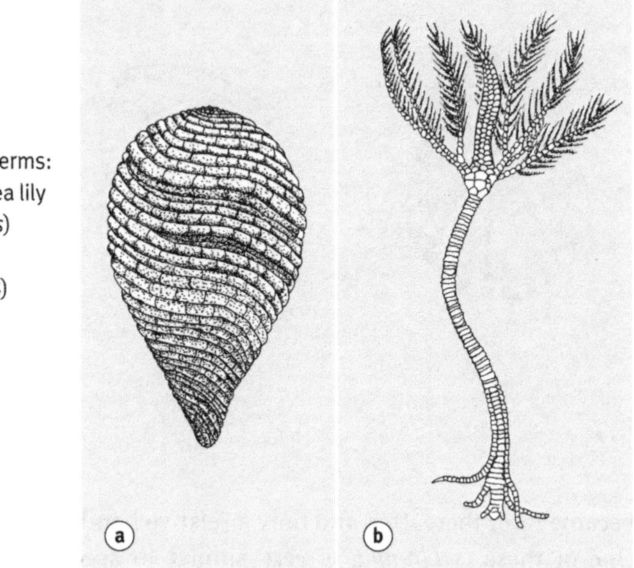

FIG. 5.11 Early echinoderms: (a) helicoplacoid; (b) sea lily (crinoid) (*Dictenocrinus*) ((a) after Fortey, 1982; (b) after Clarkson, 1998)

high. Secondly, really substantial reefs that grow to be close to the sea's surface develop only in tropical and subtropical seas. These are termed framework reefs because the growth form of an organism is responsible for providing the structure. The worldwide loss of this type of reef is often considered a sign of global cooling. Reefs consist largely of the skeletons of reef-building organisms (animals of several groups and calcareous algae) and of limestone and other particles trapped in the complex structure.

The fossil record of reef formation is a very diverse one; a variety of organisms have been the dominant reef builders during different periods. The first to appear were the stromatolite-building cyanobacteria, which could form mounds several metres in height. At the beginning of the Cambrian they were joined by the stony sponge-like archaeocyathids (FIG. 5.4), these were generally about five centimetres high, though some could be more than three times this size. Unlike modern corals they did not form true colonies, but the individuals were bound together by outgrowths of their cups (theca); there is evidence of possible competition, with some individuals overgrowing others. After about 20 million years these animal reef builders effectively died out.

In the middle of the Ordovician, some 55 million years later, reefs appeared again (meanwhile the cyanobacteria had continued to form stromatolites). This reef community was a rather complex one involving a red alga with a calcareous covering (a coralline alga), stony sponges (stromatoporoids), two types of coral (rugose and tabulate) and various moss animals (Bryozoa). After about 33 million years, at the end of the Ordovician, reefs became much restricted and many families of corals and bryozoa appear to have become extinct. The possible reasons for this are discussed later (p.62).

Weird fauna from the Burgess Shale and other sites

The animals already discussed had hard skeletons and so were relatively readily fossilized; detail has been preserved and this has been sufficient to assign them to groups of animals still living. However animals that were largely soft-bodied are seldom preserved whole, though some small mineralized part might be fossilized. Often the form of the complete animal remains a mystery until some lucky discovery is made. The conodonts are an excellent example of this; they are found abundantly from the Cambrian to the Triassic. It was not until the 1980s when Euan Clarkson found some fossils near Edinburgh, that it could be seen that they were the equivalent of teeth from an eel-like creature (FIG. 5.12) that belonged to the chordates (the group which includes vertebrates). When the relationships of fossils are unclear they are often referred to as 'Problematica', which truly reflects their status. There is considerable controversy as to whether all Problematica will, if more detail becomes available, turn out to be allied to existing animals or at least close to their ancestral line, or whether they represent dead-ends, designs that have failed, probably through chance. The first view is held by Simon Conway Morris and expounded in his book *The crucible of creation*, the second is advanced by Stephen Jay Gould in *Wonderful life*. The dispute can be crudely summarized by the two evolutionary trees in FIG. 5.13. The Gould tree has many major branches that

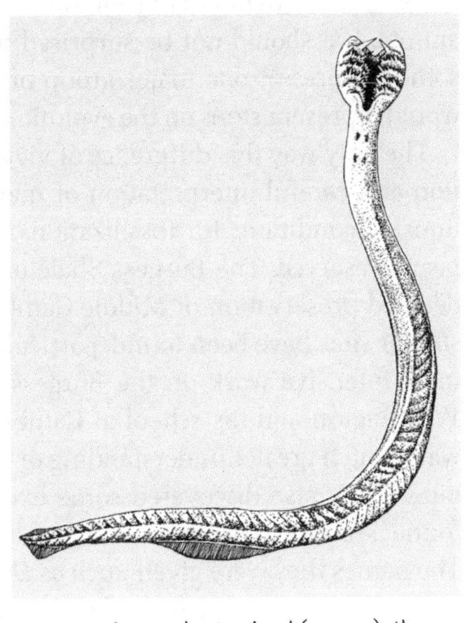

FIG. 5.12 A conodont animal (55 mm); the 'teeth' alone were termed conodonts (after Hoffman and Nitecki, 1986)

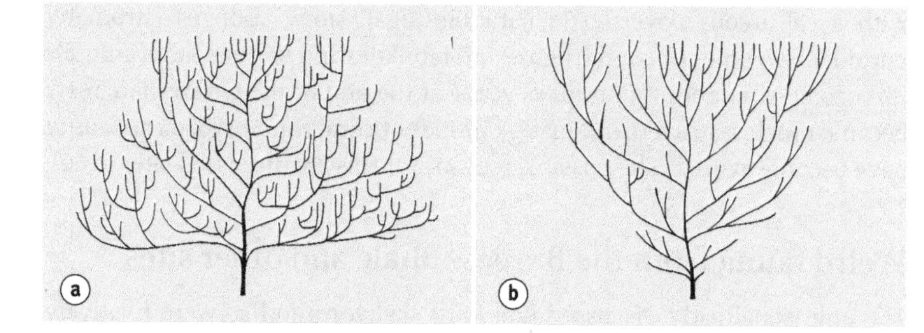

FIG. 5.13 Evolutionary trees according to two different interpretations: (a) S. J. Gould's view; (b) S. Conway Morris' view

peter out early in evolutionary history without any relatives surviving; not, he would argue, because they were bad designs, but as a result of some sudden environmental change. The Conway Morris tree indicates that most, if not all, of the Problematica were at the most short side branches from the main streams of evolution. If, as we believe, worms (Annelida), shellfish (Mollusca), and lampshells (Brachiopoda) all had, at one time, a common ancestor then we might hope to find evidence in the fossil record of such an animal. We should not be surprised to discover fossils that seem to have some features of one major group of animals and some of another. They would represent stops on the evolutionary road rather than dead-ends.

The only way this difference of views can be resolved is by the examination and careful interpretation of material from sites where, under some unusual conditions for fossilization, at least some of the soft tissues have been preserved. The Burgess Shale of British Columbia is famous for the detailed preservation of Middle Cambrian fossils and more recently other similar sites have been found, particularly Chengjiang in South China. The most intensive work on the Burgess Shale fauna was done by Harry B. Whittington and his school at Cambridge. An early fruit of their labours was a much greater understanding of the anatomy and functioning of trilobites. They also discovered some extraordinary fossils which, when first found, seemed to be so weird that they were assigned to the Problematica. The names they were given, such as *Hallucigenia* and *Anomalocaris*, reflected the puzzlement of their discoverers. Now further study of additional material from several locations has shown that these fossils can be fitted into the existing classification, though often in unique positions.

Quite commonly found fossils include certain minute calcareous plates; examples of one type have been called halkieriids. They were clearly part of something, but of what? The answer came when John Peel and Simon Conway Morris found some remarkable specimens in North Greenland that looked superficially like an armoured slug; on their backs were hundreds of these small plates arranged like tiles on a roof. At either end of the body were two shells, of which at least one looked very like that of a lampshell (FIG.5.14(a)). However the tiny plates can be regarded as similar to the skin plates or sclerites of polychaete worms. If, as is not unreasonable and is supported by molecular evidence, we consider lampshells and worms to have a common ancestor—a 'missing link'—the halkieriids are it. In the Burgess Shale another curious fossil was found and named *Wiwaxia.* Its body was also covered with small plates, though some of these were long and pointed (FIG. 5.14(b)), but they were not mineralized and shells have not been found at either end of the body. The pointed plates probably served as some protection against predators, and some are broken off as if the animal had been attacked. The difference between mineralized plates and unmineralized ones does not signify much in terms of relatedness, so it has been proposed that *Wiwaxia* is related to the halkieriids but, lacking shells, it is somewhat closer than they are to polychaete worms.

A spiky back was probably as common a defence mechanism in the Cambrian and Ordovician seas as it is now and this seems also to have been the strategy of *Hallucigenia* (FIG. 5.14(c)). Originally this animal was drawn with the spikes pointing downwards, being some sort of leg, but it is now recognized that this was an upside-down view. It is still uncertain which end is the front and which the back. The tube-like projections are the legs and such tubular legs are known as lobopods. Today two groups posses them: the minute water bears (Tardigrada) and the velvet worms (Onychophora). The former are principally found living in water films on mosses and elsewhere from the polar regions to the tropics. Velvet worms are much less common; they also live in habitats with high humidity, particularly amongst fallen leaves in tropical and subtropical rain forests, but two species are found in caves. Modern studies suggest that these lobopodians are somewhere on the ancestral line between arthropods and polychaete worms. It seems that in the Cambrian there were many lobopodians living in the seas, not only *Hallucigenia,* but also *Aysheaia* (FIG. 5.14(d)), which looks very like a modern velvet worm, and others found especially in the fossil beds at Chengjiang.

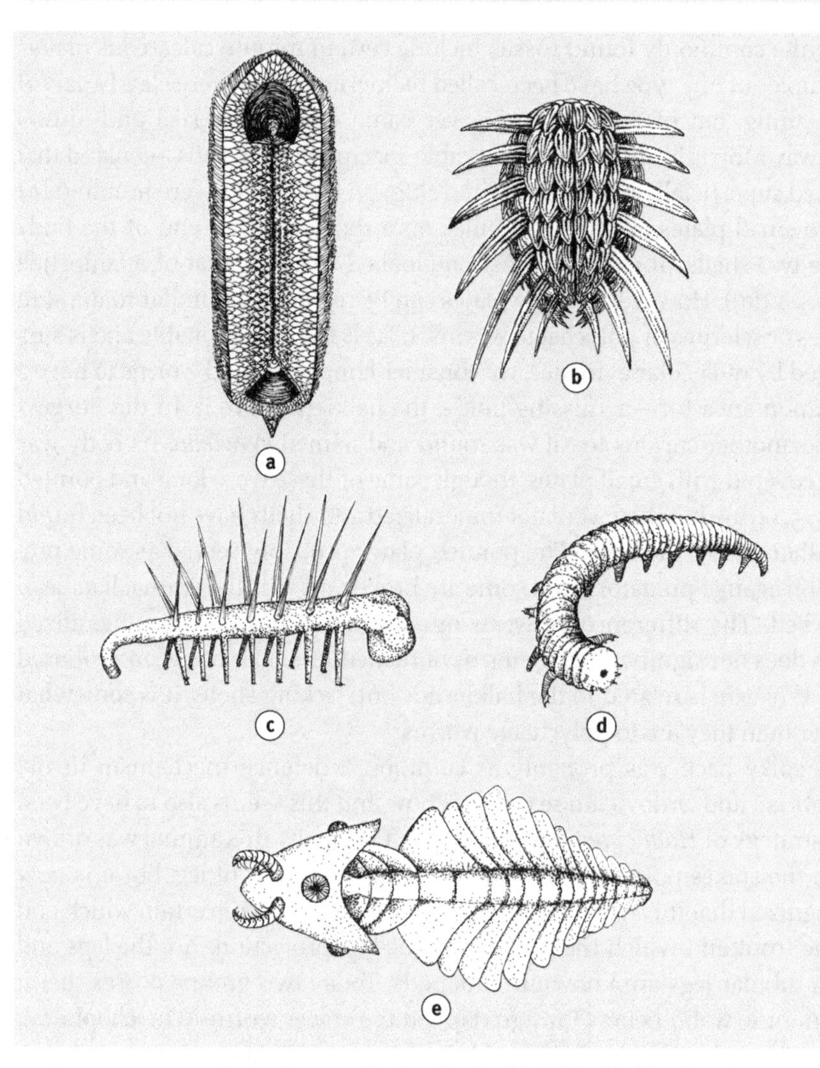

FIG. 5.14 Some Burgess Shale animals: (a) Halkierid; (b) *Wiwaxia* (c) *Hallucigenia*; (d) *Aysheaia*; (e) *Anomalocaris* ((a) and (c) after Conway Morris, 1998; (d) and (e) after Clarkson, 1998; (b) after Hoffman and Nitecki,1986)

Even the remarkable *Anomalocaris* (FIG. 5.14(e)) can be seen to represent an early stage in the evolution of one of the main existing groups—the arthropods. It was clearly a predator: the fearful pair of jointed appendages at the front would have been used to capture prey. The flap-like extensions to the body may represent the precursor of the upper part of the characteristic arthropod limb, as found in trilobites. These fossils are now not so unique: other specimens are being found that share some of their features, but which have other structures in common with existing groups. Some fossils of animals, apparently close to *Anomalocaris*, have limbs rather like those of a lobopodian (velvet worm); therefore rather than being on an evolutionary byway, these animals are somewhere close to the basic arthropod route between lobopodians and arthropods.

The evolutionary picture developed from studies of modern groups—arthropods (millipedes, crustaceans, scorpions, etc.), lobopodians, annelids and molluscans—postulates that there would have been these early stages and, if we speculate on how they would look, one comes up with animals sharing features with *Wiwaxia, Hallucigenia, Anomalocaris* and the others. Therefore it can been argued that these fossil faunas show some of the roots of modern invertebrate groups. What of the origins of vertebrates?

Vertebrate relations and ancestors

Vertebrates, the familiar fish, amphibians, reptiles, birds, and mammals, share certain features with other less well-known animals such as the sea squirts (Urochordata). The key feature, the one that gives them all their name—Chordata—is the possession, either in the young condition or throughout life, of a notochord. This is a rod-like structure that lies below the dorsal nerve cord. Another characteristic is that the wall of the pharynx, that part of the gut that comes just after the mouth cavity, has a series of slits, often termed gill slits or clefts. As with the notochord, in higher vertebrates these are found only in the embryonic stages. These gill slits also occur in arrow worms and related animals, but these do not have a true notochord and are placed in the appropriately named and related group Hemichordata. At this point (around 540 Mya) in the evolutionary story we would postulate the existence of hemichordates and primitive, perhaps worm-like chordates.

Black shales deposited in this period contain fossils that look like faint pencil lines (FIG. 5.15), these are called graptolites; but in spite of their abundance, from the Ordovician to Devonian, there has been uncertainty until recently as to their exact zoological affinities and mode of life. The discovery in 1989, at a depth of 250 metres in the ocean off New Caledonia of what has been claimed to be a living fossil enables us to have a much clearer understanding of graptolites and how they lived. This finding confirms that they are hemichordates and belong to the Pterobranchia. The fossils represent the material secreted by the animals, the individuals of which are

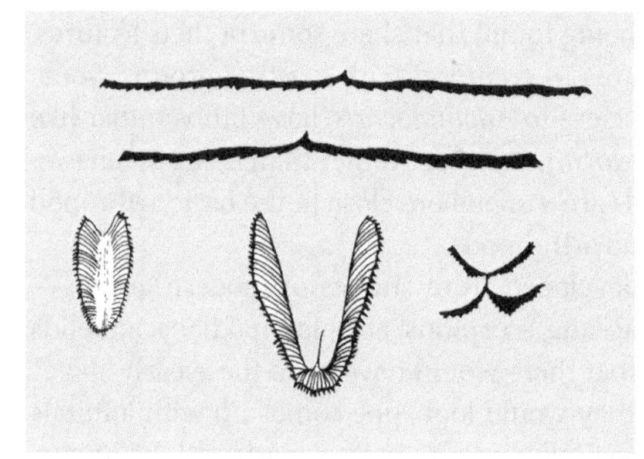

FIG. 5.15 Graptolite fossils (after Fortey, 1982)

termed zooids; we can now see that during fossilization the 'house' has remained, but no sign of the soft-bodied occupants survived. Each graptolite represents the home of a colony of zooids and each lives in a sac-like tube; these are cemented together, the entrance to each tube being the projections on one side of the graptolite. At the end of the colony there is often a spine. It is now known that the zooids leave their individual tubes and crawl along the colony and up the spine where they feed, filtering particles from the water with their tentacles (FIG. 5.16). During this excursion they secrete additional material and so add slightly to the height of the spine. When graptolites are discovered they are often found in large numbers, but usually there are not many other types of fossil. From this observation and the black shales in which they generally occur we can conclude that many graptolites lived, like their 'descendant', in waters that were deep in the oceans though not necessarily on the bottom. Modern pterobranchs are simpler in their colonial structure than most of the graptolites, some of which have a structure that is interpreted as a float and were probably planktonic in the upper waters of oceans.

Mention has already been made of the conodonts (p. 55), tooth-like fossils, often with a complex structure of many cusps found, after long years of speculation, to belong to an eel-like creature (FIG. 5.12). Usually this is about five centimetres in length, but a fossil eight times this size has recently been found. The exact affinities of conodonts are still uncertain, but it is generally agreed that these lie with the primitive chordates.

Fossils of what are believed to be genuine chordates have been found in the Lower Cambrian deposits at Chengjiang, China and in the Middle Cambrian Burgess Shale. *Cathymyrus*, as its name implies from China, shows evidence of gill slits and has V-shaped blocks of muscle in the body

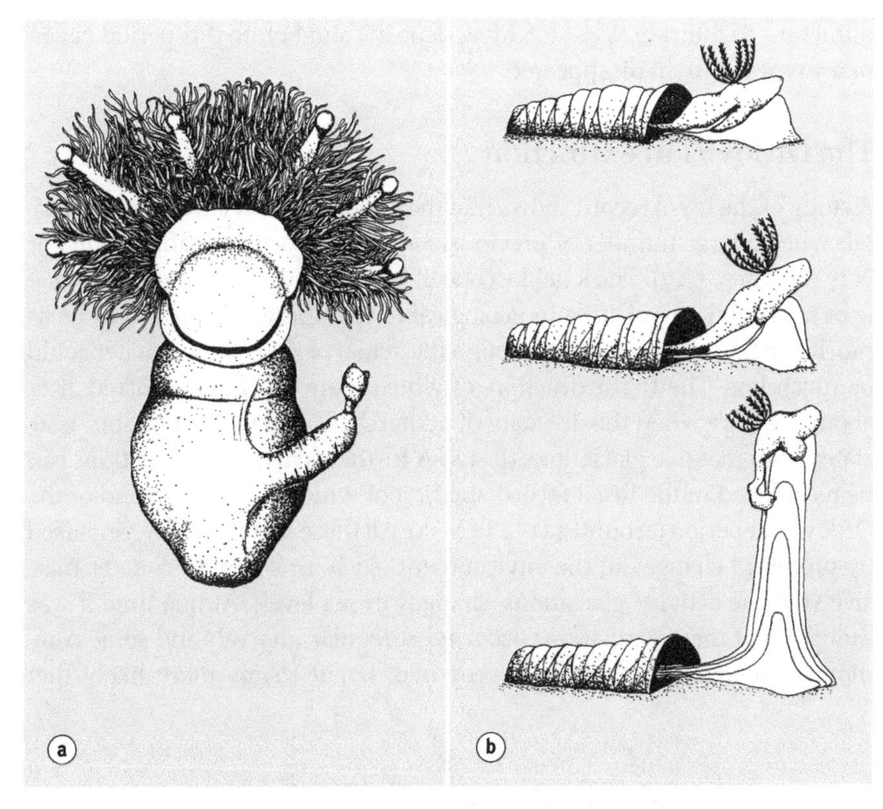

FIG. 5.16 Living pterobranchs (*Cephalodiscus*): (a) a single animal; (b) diagram suggesting the mechanism for the secretion of spines by the animals during a series of excursions from their tubes. ((a) after Parker and Haswell, 1897, from McIntosh; (b) after Dilly, 1993)

wall and other features that place it in the most primitive group of chordates, the Cephalochordata. This includes amphioxus, a much studied elongate animal found today in the sand, at small depths, in many temperate seas. In the Burgess Shale a rather similar fossil, known as *Pikaia*, has been found, but the gill slits are not visible. We can expect further work on these and other, yet-to-be-found, fossils from Chengjiang and other sites to throw more light on to the structure and habits of animals at the base of the chordate–hemichordate evolutionary branch. How exciting these are likely to prove is shown by the discovery in 1999 in the Chengjiang beds of two true chordates, indeed fossil fish, *Myllokummingia* and *Haikouichthys*. They have been placed in the group Agnatha (p. 69) the members of which are numerous in Silurian (438–408 Mya) deposits. But before this period began many types of fossil disappeared.

The Ordovician extinction

A study of the fossil record shows that there were some relatively short periods when a large number of previously abundant forms became extinct or very rare (FIG. 5.17). The kaleidoscope of life was given a strong shake. One is of course talking of time in geological terms: these periods might be as much as a few million years, although the actual period of the disaster could be much less. The first extinction, of which there is a hint, occurred after about 900 Mya when the diversity of acritarchs began to fall, possibly associated with massive glaciations (p. 32). A further five major extinctions can be recognized in the fossil record, the first of which came at the end of the Ordovician period (around 441–438 Mya). All these extinctions were caused by profound changes in the environment (such as asteroids, comets, massive volcanic activity, glaciations, changes in sea level). At one time it was thought that these extinctions occurred at regular intervals and some common extraterrestrial cause was proposed, but it seems more likely that

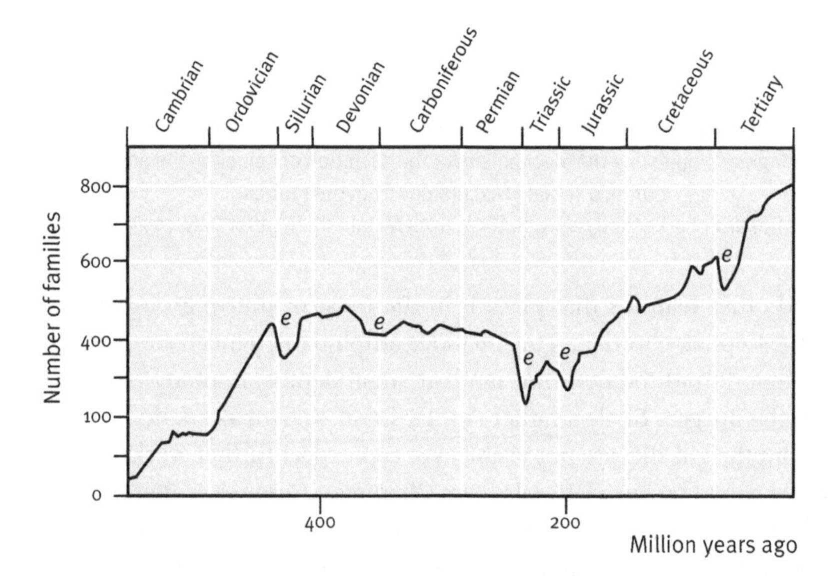

FIG. 5.17 The increasing diversity of marine animal types as represented by the number of families throughout the fossil record, with major extinctions marked 'e'. (after Sepkoski, 1981)

different causes operated at different times and these possibilities will be reviewed at the end of each chapter.

Animals that became extinct were not those that were badly adapted; indeed they were likely to be those most closely adapted to the existing environment, but unable to adjust to the challenge of the changes. Some of those that survived did so because of flexibility, some because the new environment was, by chance, not so adverse for them; and others, also by chance, because they lived in a part of the environment relatively unaffected by the environmental changes. Extinctions are not limited to these periods of mass extinction; they occur all the time, but on a much smaller scale. Again they are commonly due to some change in the environment, but in these cases the environmental change may be biological: some other organism adversely affecting, directly or indirectly, the species that became extinct. The importance of this mechanism, competitive displacement, is a matter of controversy but it certainly occurs today, as witnessed by the manner in which goats, rats and cats, introduced on oceanic islands by sailors or settlers in the last few centuries, have driven many native animals to extinction. In general, when two faunas mix there is a reduction in the total number of species; the effect of the opening up of the land bridge between the North and South Americas and the mixing of their faunas is one example in the fossil record. Competitive displacement may also have been important in organisms that compete directly for space, principally plants, but also non-mobile animals such as barnacles. So we can envisage the kaleidoscope of life being tapped all the time and occasionally shaken, more or less vigorously, giving a whole spectrum of rates of change. The most vigorous shakings are the mass extinctions that provide a general framework within which the chapters of this book are mostly organized.

Twelve groups that were conspicuous members of the fossil assemblages of the Cambrian and Ordovician became much reduced at the end of the latter period. For example about half the genera of lampshells, cephalopods and graptolites disappeared from the fossil record and higher proportions of trilobites and conodonts were lost. At this time Gondwana was moving as a single continent towards the South Pole, triggering a period of extreme glaciation. The formation of ice sheets on continents leads to a marked falling of the sea level. However recent work has indicated that the Ordovician extinction was a series of events. The first, and more minor, phase of extinction may have been due to some of the components of the northern supercontinent, Laurasia, coming closer together. The faunas became less local in character (provincial) and more similar within each individual group of tectonic plates: that containing North America and Europe on the one hand and that containing various parts of Asia on the other. This is considered to represent a case where faunal mixing caused an overall reduction in total diversity. The mixing is thought to have been caused by the carriage of the planktonic larval stages across the narrowing intervening seas.

The main phase of the reduction in diversity is generally attributed to the combined effect of the lowering of global temperatures and the falling of sea levels following the extensive glaciation of Gondwana; what is today the Saharan desert was at this time at the South Pole (FIG. 6.1). Even faunas not directly affected by the glaciation would have been influenced by the cool climate and by the draining of the shallow seas that lay on continental land masses (epicontinental seas).

The last phase arose when the Gondwana ice sheet melted relatively rapidly—for reasons not fully understood—and the sea levels increased. This generated anoxic conditions that in turn led to the deposition of black shales. These shales mark the advent of the next era, the Silurian.

CHAPTER 6

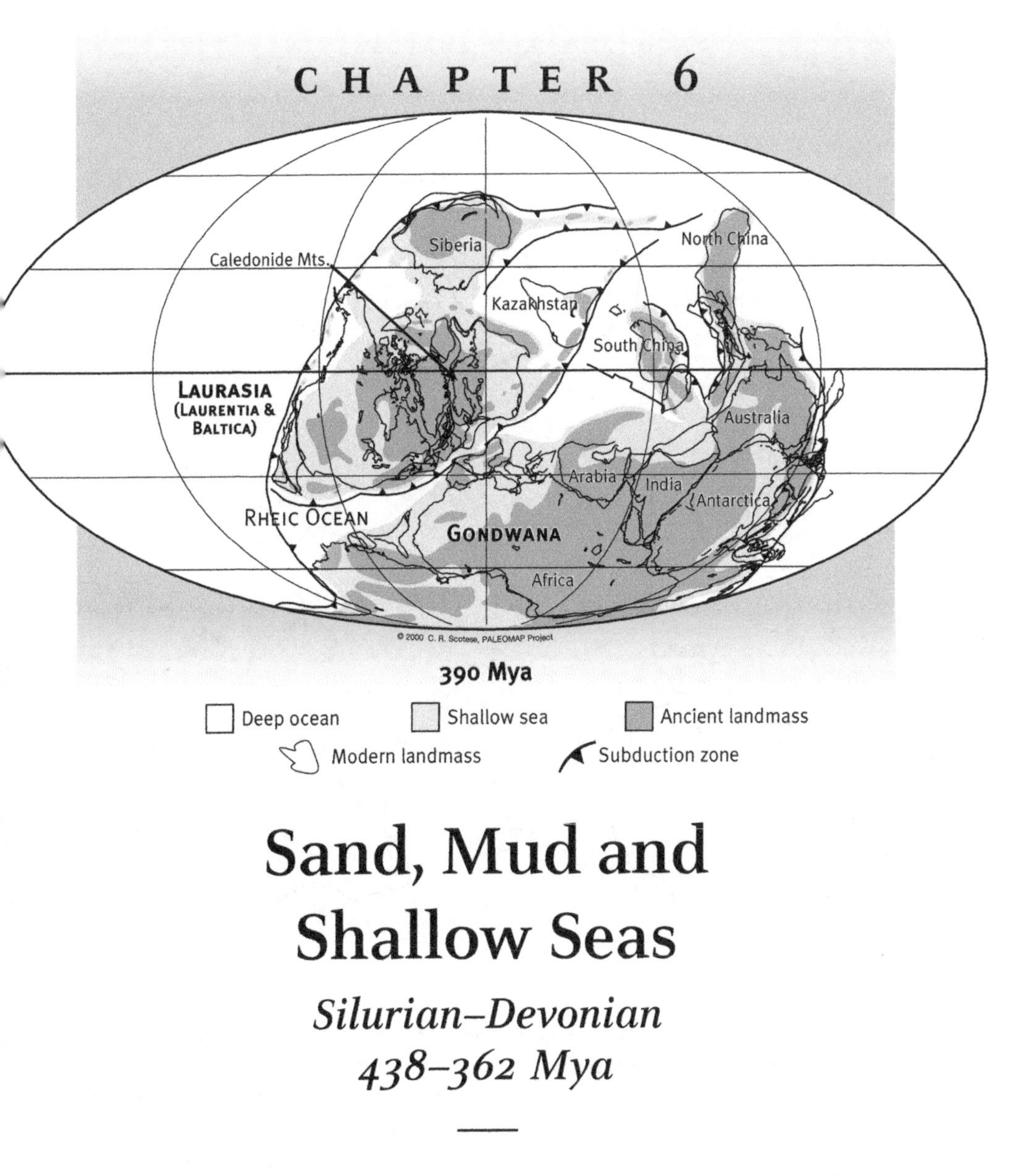

Sand, Mud and Shallow Seas

Silurian–Devonian 438–362 Mya

DURING this period—the Silurian and Devonian—the continental land masses were, in the main, slowly approaching each other. Gondwana was a single entity, moving over the South Pole. So the continental land masses that now constitute Central Africa, Southern Africa, and South America all passed over the pole until Gondwana was clear of the pole, only to move back again at the end of the period (FIG. 6.1). The northern edge of Gondwana was fringed by extensive shallow seas and at one stage these

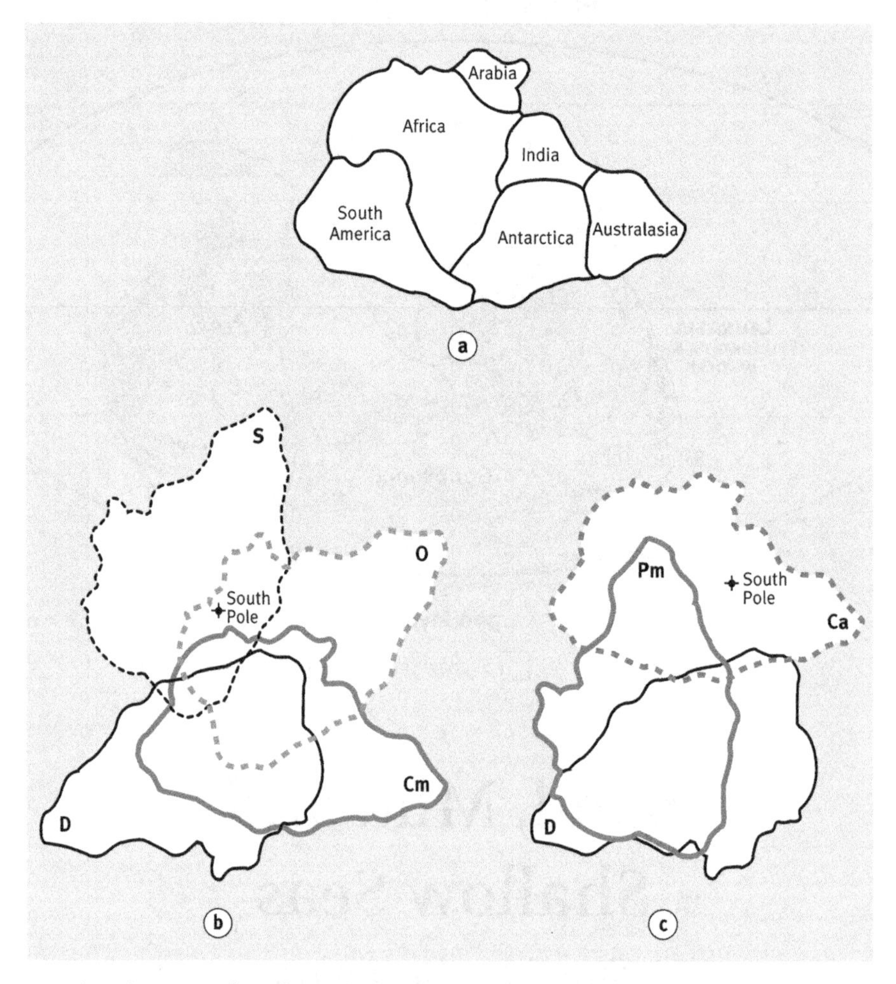

FIG. 6.1 Diagram to show the positions of Gondwana relative to the South Pole during six successive geological periods (Cambrian—Permian). The position is approximately that of the middle of the period. The shape of Gondwana was not constant over these 250 million years: (a) Gondwana as represented showing relationship to present land masses; (b) movement over Cambrian (Cm), Ordovician (O), Silurian (S), and Devonian (D) periods; (c) movement over Devonian (D), Carboniferous (Ca), and Permian (P) periods

were contiguous with those of the land mass that now includes China—one of several land masses in the north, that would come together to form the supercontinent Laurasia. The North American and Northern European (Baltica) continents came together, forcing up mountain ranges such as the Appalachians and the 'Caledonide' of Scotland and Norway. In the Devonian these young mountains were eroded, the rivers carrying abundant sediment into vast deltas where it ultimately settled to form the rock known as Old Red Sandstone. Sea levels were particularly high during the Silurian, when two-thirds of what is today North America was under shallow seas. The northern European and North American land masses, often called the 'old red sandstone continent', sat more or less on the equator. Over this period Gondwana and all the components of Laurasia were moving together, thus narrowing the oceans between them.

With the earth constituted in this way, many of the organisms found in the fossil record are very likely to have dwelt in warm shallow seas (FIG. 6.2), lagoons, deltas and even flood plains, for there is evidence of seasonal droughts and floods. There has been some debate as to whether the sites where fossils have been found were originally fresh or salt water. This can now be determined by the ratio of two isotopes (p. 8) of strontium in the associated deposits. However in reality, as in deltas today, the salinity of the water would have often varied, both over short distances and over time. With this knowledge of the conditions on earth and the habitats available one might expect fish to have been prominent—and so they are: this has been called 'the age of fish'.

The radiation of fish

It is apparent from the history of life so far that the rate of evolution is not constant throughout time; there was the Cambrian 'explosion' and the Ordovician 'extinction'. We have noted how an 'explosion' often follows a 'fuse period' when important evolutionary developments take place, like actors taking their positions backstage before the curtain rises. Such explosions, less dramatically termed 'radiations'—great increases in the species of one group, occur throughout the fossil record, more particularly after periods of extinction. At one time it was thought that groups that radiated were those that had out-competed their less well-adapted predecessors. However, a more detailed study of the fossil record shows that changes in the environment had frequently led to the demise of one group many years before the second group started to become dominant. In general it seems that radiations happened when a group of organisms had adaptations that enabled them to occupy many vacant niches (metaphorically,'ways of earning a living'). Early in evolutionary history one can envisage the occupation of niches that had never before been occupied. This seems to apply in general to the radiation of the fish at this period. Of the groups severely affected

FIG. 6.2 Silurian sea, showing agnathan and acanthodian fish, nautiloid, lampshells, trilobites, jellyfish, sea lilies (crinoids), rugose corals, starfish , sea anemone, sea scorpion, shellfish (molluscans), worms and a shoal of conodont animals (top right)

by the Ordovician extinction only the cephalopod molluscs might have occupied similar niches before and after the extinction.

Many of the different body plans that evolve in a radiation eventually become extinct. It is interesting that, of the fifteen or so major groups of fish that evolved around this period, only five have members living today and three of these contain only a handful of species—unless one includes all four-legged vertebrates (tetrapods) whose ancestry can be traced to one of these groups of fish. Apart from the cartilaginous fish (sharks, rays, etc.) almost all living fish are members of one subgroup (Teleostei) that represents but one of thirty-four different body plans evolved at various times within one of the fifteen major groups just mentioned.

Jawless fish

The first fish to appear in the fossil record lacked jaws and are often termed the Agnatha. Recent finds suggest that they may have evolved as early as the Cambrian (p. 59) and two types have been found in the Ordovician. From the beginning of the Silurian Agnathans become frequent in the fossil record. Six different groups are recognized from the period covered in this chapter (FIG. 6.3) and many of them seem to have been abundant at various

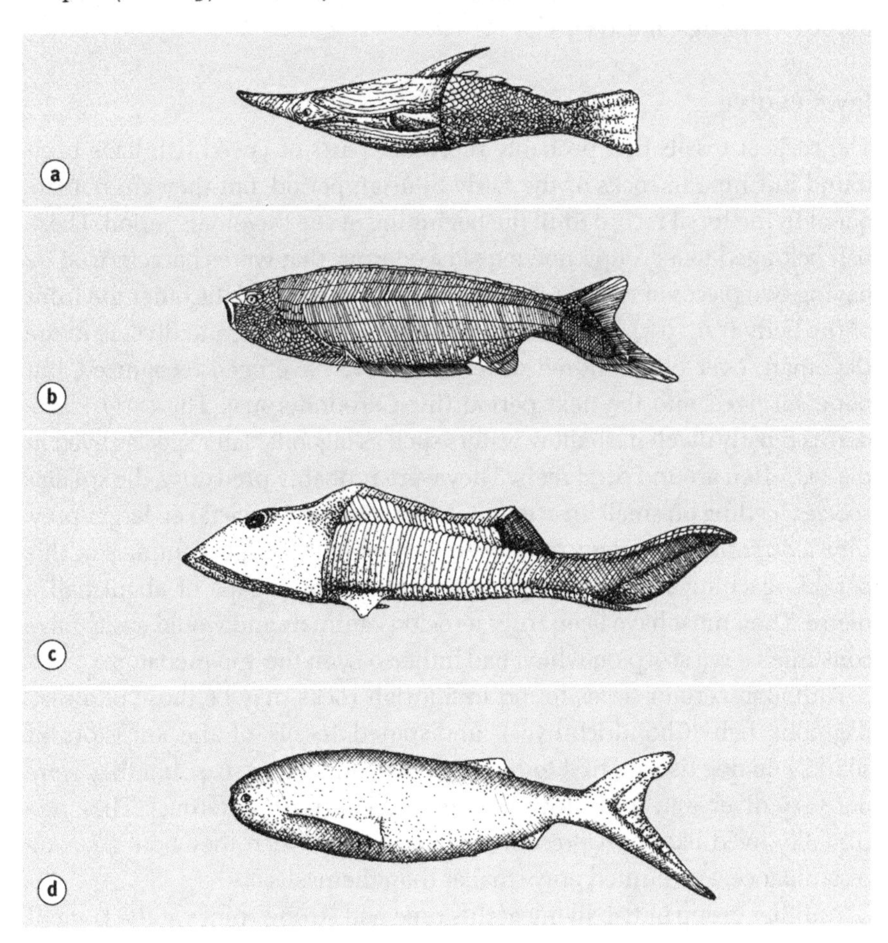

FIG. 6.3 Some jawless fish (Agnathans): (a) Heterostracii; (b) Anaspida; (c) Osteostraci; (d) Thelodonti (after Janvier, 1996)

times, but none are found after the end of the Devonian. In addition there are two Agnatha living today, the hagfish and the lamprey: first found in the Carboniferous period, they are but poorly represented in the fossil record. Although these modern species are essentially parasitic or scavenging, it is believed that most Agnathan were suspension feeders, filtering out minute food particles from a current of water. The body form of the majority, with their large armoured and flattened head shields, suggests that they lived on the mud or sand at the bottom of the sea, lake, or river, sometimes largely buried—a benthic life-style. The members of two groups (Anaspida and Thelodonti) lacked the head shield and probably lived in the upper waters of the sea—a pelagic lifestyle.

Jaws evolve

The earliest fossils that probably represent parts of jawed fish have been found in China in rocks of the Early Silurian period, but they are not frequent in the fossil record until the beginning of the Devonian period. These fish belonged to a group known as Placoderms, that were characterized by having two pieces of armour , one covering the head and the other the front of the body (FIG. 6.4). They were abundant, widespread, and diverse in the Devonian; over two hundred different genera have been recognized, but none survived into the next period (the Carboniferous). The early Placoderms mostly dwelt in shallow waters such as lagoons; later species lived in the sea, often around coral reefs. They were probably predatory, the smaller species feeding on small invertebrates, but some will have taken larger prey. *Dunkleostethus* and *Titanichthys* were probably the largest animals at this period, reaching six metres in length and having a gape of about half a metre. They must have been truly ferocious animals and would easily have consumed a sea scorpion which had hitherto been the 'top predator'.

Although certain scales found in Silurian rocks may be those of a cartilaginous fish (Chondrichthyes), undisputed fossils of the ancestors of sharks and dog fish are first found in the Devonian (FIG. 6.4), but they were not very diverse in form until the next period (Carboniferous). They presumably lived like most present day species, to which they bore a strong resemblance, and hunted prey smaller than themselves.

Another group of fish living at this time had strong spines at the front of most fins, but none have survived until the present. The other two groups that apparently evolved in this period are termed bony fish and both have

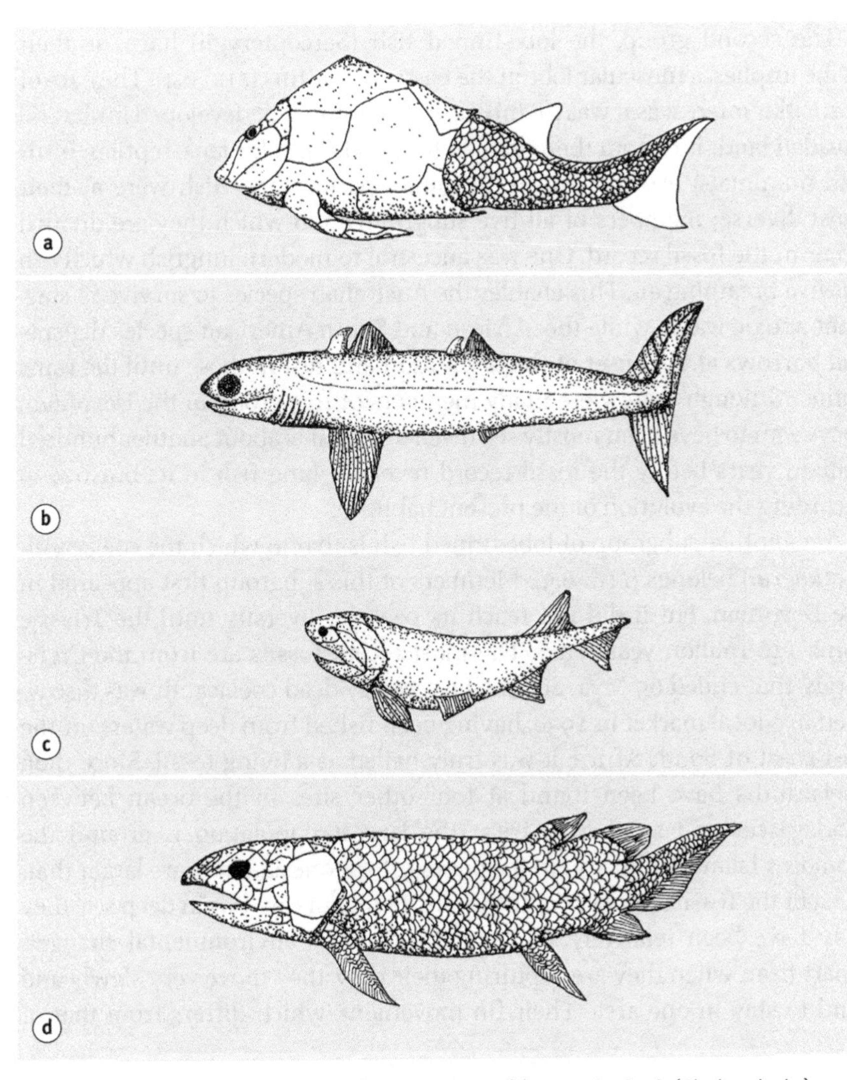

FIG. 6.4 Some Devonian fish: (a) a placoderm; (b) an early shark (*Cladoselache*); (c) an actinopterygian; (d) a lobefinned fish (*Dipterus*) ((a) after Cowen, 1995; (c) and (d) after Janvier, 1996)

many 'descendants' alive today, including ourselves! One of these groups, ray-finned fish (Actinopterygii) are typical fish (FIG. 6.4) and, as already mentioned, the great majority of fish found today are members of a division of this group, although there were not so many species in the Silurian and Devonian periods.

The second group, the lobe-finned fish (Sarcopterygii) have, as their name implies, a muscular lobe at the base of their fins (FIG. 6.4). They are of particular interest as it was members of this group that developed limbs and invaded land; it is from them that all tetrapods (amphibians, reptiles, birds and mammals) evolved. In the Devonian period these fish were at their most diverse; members of all five subgroups into which they are divided occur in the fossil record. One was ancestral to modern lungfish which can survive breathing air. This enables the Australian species to survive in stagnant anoxic water, while the African and South American species dig special burrows at the onset of the dry season and rest in these until the rains come. Although there were many members of this group in the Devonian, they seem to have been mostly sea dwellers and it is about another hundred million years before the fossil record reveals a lung fish in its burrow, so recording the evolution of the present habit.

Yet another subgroup of lobe-finned fish is that to which the coelacanth (*Latimeria*) belongs (FIG. 6.5). Members of this subgroup first appeared in the Devonian, but it did not reach its peak in diversity until the Triassic some 140 million years later. The most recent fossils are from the Cretaceous, that ended 65 Mya. So when a recently-dead coelacanth was discovered in a local market in 1938, having been fished from deep waters off the east coast of South Africa, it was truly hailed as a living fossil. Since then coelacanths have been found at four other sites in the ocean between Madagascar and southern Africa (the largest population is around the Comoros Islands), and also in the ocean off Indonesia. They are larger than most of the fossil members of the group. Living in canyons in deep sea, they may have been relatively sheltered from many environmental changes. Apart from when they are capturing their prey, they move very slowly and tend to stay in one area. Their fin movement, which differs from that of

FIG. 6.5 The coelacanth (*Latimeria*) (after Millot, 1955)

other living fish, is particularly interesting and shows the same sequence of limb movements as that of crawling amphibians and most mammals when they walk. The front fin on one side moves forward at the same time as the rear fin on the other side.

Fins to limbs

Lobe-finned fish have long been thought to be the forerunners of amphibians and other four-limbed animals (tetrapods). It was believed that some of these fish may have been at least partly amphibious and that they used their fins to drag themselves out on to the mud banks of the Middle Devonian. However recent research indicates that the fish closest to a tetrapod remained aquatic. One such fish is *Panderichthys* (FIG. 6.6) found late in the Devonian period. It lacks fins on its back as well as an anal fin. Its form is well suited for living in shallow water and forcing its way through algae and other aquatic or semiaquatic plants.

It is probable that the first tetrapods found in the fossil record, such as *Acanthostega* (FIG. 6.6), were also aquatic. This, and related species, are found in the late Devonian and still retain the tail fins of fish. However, a number of significant evolutionary steps have occurred. The paired fins have become eight-toed (front) and seven-toed (hind) limbs. Whereas the front fins of lobe-finned fish are the stronger, the longer hind limbs in *Acanthostegia* would appear to have had more leverage—this animal was

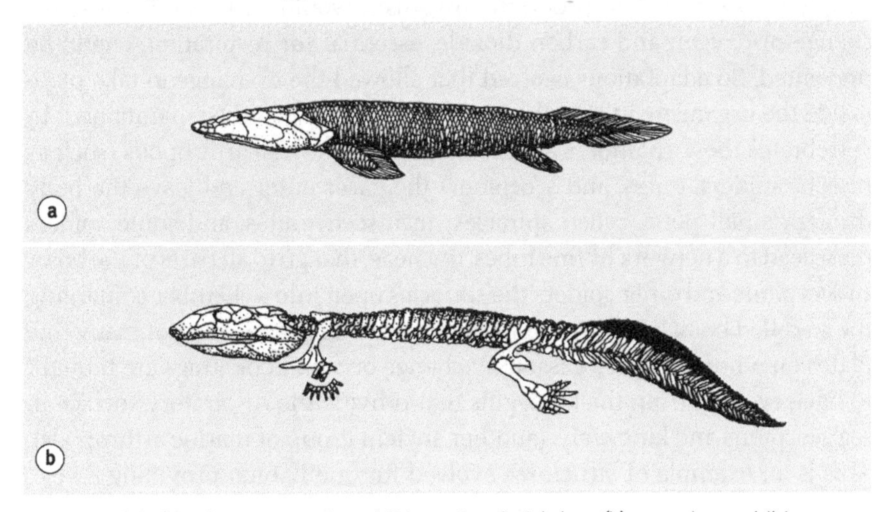

FIG. 6.6 (a) A forerunner of amphibians, *Panderichthys*; (b) an early amphibian, *Acanthostega* (after Carroll, 1995)

on the way to being 'rear wheel driven' like most tetrapods. There were also various changes in skeletal structure which conform to the tetrapod form, rather than to that of fish. So, as often in evolutionary history, it seems likely that the adaptations necessary for this key step evolved in one habitat, in this instance the aquatic one, equipping the organism for another habitat, the land. By the late Devonian, tetrapods were certainly on land. A number of species that were probably amphibious have been identified from late Devonian rocks from several parts of the world, but after this there is a twenty-million-year gap in the fossil record.

The challenge of land

What challenges did land present to organisms that evolved in the sea? The fundamental difference lies in the properties of air and of water. Life started in water, which screens out harmful ultraviolet radiation. But until the level of oxygen was sufficiently high for an ozone shield to build up, the rays would have been fatal to an organism living in air, on the land surface. This would have delayed the colonization of land.

A major obstacle to life on land is the risk of dehydration, so that the water-mediated processes on which life depends can no longer function. As one can observe with a jellyfish stranded on a beach, without appropriate adaptations marine organisms soon dry out and die on land. If the skin were simply to become waterproof, the organism would suffocate because exchange of oxygen and carbon dioxide, essential for respiration, would be prevented. So adaptations evolved that allowed the exchange to take place inside the organism, in chambers where humidity could be maintained. In vertebrates these chambers are the lungs. In terrestrial arthropods (such as insects, spiders, mites, and scorpions) the gases enter and leave the body through small pores called spiracles; in insects, mites, and some spiders these lead to a network of fine tubes, tracheae, that go to all parts of the body. In scorpions and other spiders the spiracles open into a chamber containing the so-called book-lungs. These, as their name suggests, consist of many fine plates on whose surfaces gaseous exchange occurs. Book-lungs are thought to have evolved from the book-gills that provide the respiratory surface in sea scorpions and king crabs (another ancient group of marine arthropods). This is an example of structures evolved for one habitat providing an element of pre-adaptation to another. It was suggested above that the evolution of legs from fins may have been another instance of this. In plants there are pores (stomata), on the leaves and stems that can be opened and shut. With these adaptations for respiration, the surfaces of terrestrial organisms can be made impermeable with waxes and greases. In general, the drier the habitat the more waterproof the surface.

Animals that invaded land already had an internal circulatory system—the blood system—used for carrying respiratory gases to and from their water bathed-gills, as well as transporting nutrients and waste products. In many terrestrial animals these systems are linked to the newly evolved or modified enclosed respiratory surfaces and function as before, but the blood system is of little importance for the transport of respiratory gases in insects and others with tracheae.

Water supplies still need to be replenished and this is generally done by drinking or in plants, through roots. There are special problems for eggs and seeds that cannot do this and they have a wide range of adaptations for avoiding dehydration. Amphibians return to the water to lay their eggs; reptiles and birds provision their relatively short-lasting egg stages with ample fluid; while most mammals retain their young in a 'pond' of fluid in the mother. The eggs of many insects are laid in water, soil, or moist leaf litter. Those laid in more exposed situations have a variety of waterproofing mechanisms and often hatch quickly. Longer lasting egg stages are frequently laid on, or even in, plants which no doubt provide some humidity. There is virtually no metabolism in plant seeds until they become wet.

There are animals such as earthworms and eel worms that have few if any of these adaptations but which appear to live on land. In fact they depend on the water films that usually exist in the soil or in the fluid interiors of plants and animals. In dry conditions earthworms burrow deeply and in a small chamber will tie themselves in a tight knot, so reducing the surface area exposed. Slugs and snails cannot move far on a surface unless it is covered by a thin film of water.

Higher plants have evolved vessels to carry water from the roots, and food from the leaves. These are readily seen as veins in most leaves and they are grouped inside stems and trunks. They are found in their most primitive form in mosses. Though mosses are very resistant to drying, they require moist conditions for growth and, like other plants early in the evolutionary tree, depend on water films for the movement of their sex cells during reproduction.

In terms of the energy used in animal metabolism, the cheapest nitrogenous waste product is ammonia, but this is toxic and so has to be quickly 'washed out' of the animal. It is commonly the waste product of aquatic animals. At some metabolic cost it can be transformed to urea and, with more cost, this can be converted to uric acid. The latter may be stored in a crystalline form, which is thus the least toxic but most expensive option. Which of these three waste products an animal produces depends very much on the availability of water. As they do not engage in muscular activity plants have fewer toxic waste products, but those that they do produce are often turned into complex substances stored in special reservoirs in the plant. The storage of these so-called secondary plant substances also serves another purpose as they often function as chemical defences against herbivores.

In water the non-mineralized tissues of an organism are effectively weightless. Protoplasm, the matrix of the cell, has approximately the same specific gravity as sea water. That is why we can float in sea water (the air in our lungs effectively balancing the weight of our bones). In the air, however, gravity exerts its pull and the organism has what is termed the problem of self-weight. It must have a skeleton to hold the various parts of its body in position. Those animal groups that have very successfully invaded land are the two that had jointed mineralized skeletons—arthropods and vertebrates. This is yet another example of the way in which pre-existing adaptations have proved useable, indeed essential, in a new environment. Large organisms can live in the sea with relatively light skeletons; the giant kelp seaweed (*Macrocystis*) can reach over fifty metres in length yet its stem is a fraction of the trunk of a redwood tree of the same height. Giant squids, which have no real mineral skeleton, are nearly the weight of an elephant. Although whales have an internal skeleton this is insufficient to support their weight on land; left for long, beached whales virtually crush themselves. The problem of self-weight becomes more severe the larger the animal, for weight is proportional to volume and therefore increases in proportion to the cube of the length. But the strength of a piece of skeleton is proportional to its cross-section and thus increases only in proportion to the square of the length. So larger animals need proportionally thicker and heavier skeletons. But the thicker the skeleton the more muscle required to move it and so the animal becomes heavier still. This balance probably determines an upper limit to the size of animals on land and is the reason why large animals move their limbs relatively slowly and tend to keep them straight, thus reducing any bending strain on the bones. Their stride is long, allowing them to cover the ground fairly quickly. It is therefore not surprising that the largest known animal, the blue whale, which may weigh as much as one hundred and twenty tons, is marine; whereas the largest dinosaurs are usually estimated to have weighed about seventy tons, and a large bull African elephant no more than six and half tons.

We have already noted that armour and speed are two alternative strategies for defence against predators. Whereas some marine organisms, such as the squid, have gone for speed, others have evolved armour. The buoyancy of the sea makes it possible for marine animals to have very heavy armour and still be relatively agile. The ammonites (p. 130), the largest of which was over two metres in diameter, and the present day *Nautilus*, both belong to the same group (cephalopod molluscs) as the speedy squids, but they have retained and even strengthened their armour, the molluscan shell.

A further difference between terrestrial and aquatic habitats arises because water filters light, so that it decreases to zero at a certain depth, depending upon the particles present in the water. So in the ocean the best place for a photosynthetic organism to be is close to the sunlight. The rate of photosynthesis is highest near the surface and at the surface. Photosynthesizing bacteria and protists can stay in this position, so forming a major component of the plankton, and consequently reducing the light available at lower levels. On land the rate of photosynthesis does not change materially with height above ground, but a plant that is taller will obtain the most light and shade out those below. Once on land, photosynthesizing organisms therefore evolved to out-compete each other by growing taller, but this required the development of all the structures we find in trees—in other words the evolution of the arborescent form. Hence microscopic organisms have continued to be the primary producers of the food chain of the oceans, whilst in most places on land this role is fulfilled by trees.

Many marine or aquatic organisms actually spend their whole lives in the water, unattached to a rock or any other substrate. By contrast, in the terrestrial environment all organisms are actually or metaphorically 'rooted' to the soil or another organism, while the air serves simply as a medium through which they can move quickly, by flight or wind transport or a mixture of both. The density (numbers per volume) of organisms in water is usually much greater than in the air; this makes filter feeding (the trapping of other organisms or their products as they pass across the body) a viable life-style. Many of the marine organisms mentioned in earlier chapters are filter feeders. Some, like sea lilies and corals simply extend their tentacles; others like mussels and barnacles create modest currents across their food-trapping organs. Not surprisingly these filter-feeders did not become adapted to terrestrial life. Indeed the only terrestrial animals to have adopted this mode of life are web-building spiders. They have compensated for the low density of food particles in the air by greatly extending their body in the form of the web and capturing, in comparison with marine animals, large particles. An orb spider builds a web that commonly covers two thousand times the area spanned by its outstretched legs!

In marine and aquatic habitats filter feeding results in there being many sessile animals that often compete for space, like mussels and barnacles on shoreline rocks. In terrestrial environments few animals can obtain their food by waiting; movement is essential even in ambush hunters, and there are therefore virtually no free-living immobile animals on land. The only exceptions are certain plant-feeding insects, gall dwellers and scale insects that depend on the plant bringing them their food in its sap. Parasites, especially internal ones, often have relatively immobile stages, but they too depend on the flow of a fluid— that of their host. On land therefore, in contrast to much of the seabed, it is the photosynthesizers, the plants, that compete for surface space.

Apart from organic food particles, water carries mineral nutrients. Photosynthesizing organisms living in water can therefore absorb minerals they require over their whole surface. On land these mainly come from the soil, which is another reason for an internal transport system. Plants that live on other plants (epiphytes, like bromeliads) do not have access to soil and often depend on insects to bring them their mineral nutrients, especially nitrogen. They either trap these insects or have a special symbiotic relationship with them; for example, evolving complex chambers in which certain ants live.

The last major difference between land and watery habitats is in the range and speed of fluctuations in temperature. This is due to the high specific heat of water in contrast to the low value for air. We experience this every sunny day at the seaside. During the day the sea will be cool (or even cold), but the air will be warm, yet both have had the same amount of sunshine. At night the sea will have lost very little heat so it may seem warm in comparison with the air which has cooled down quickly. Temperature ranges highlight this effect: the daily range at the top of grass on a lawn in England, on a clear summer's day, can be 30°C; this daily range on the land is nearly as much as the total range of the average temperatures of the oceans stretching from the arctic to the tropics. The annual range in sea temperatures is small: around 9°C in British seas. To summarize, on land there are much greater variations in temperature, both in space and time, than there are in the seas. Taken with the variations in rainfall patterns and in the soil, it means that different microclimates occur on land on a much smaller scale than in the seas. Put another way, the terrestrial ecosystem is in general much more patchy than the marine one; it has more and varied microhabitats.

The effect of this latter point can be seen if one compares the number of different species of multicellular organism on land, with those found in the sea (FIG. 6.7(a)). By far the greater proportion are terrestrial, largely due to the great diversity of insects and flowering plants adapted to the many different microhabitats on land. In evolutionary terms, in animals, this multiplicity of species are derived from few fundamentally different body plans, represented in FIG.6.7(b) by groups ranked as phyla; thirty four live in a watery environment, only two have most of their members living surrounded by air whilst one (chordates) is well represented in both environments. Most of the major types of body plan that evolved in animals in aquatic environments did not adapt to terrestrial life. Plants, in contrast to animals, evolved on land and all their phyla are predominately terrestrial.

Debris on the shore—a stepping stone?

Arthropods were the first group of animals to move into terrestrial habitats. Unfortunately the fossil record is at present relatively sparse, but from various deposits dated between 420 and 375 Mya members of a variety of groups have been identified. Perhaps the most surprising feature is that the majority of them can easily be placed in groups living today. The diversity of their body plans (FIG. 6.8) suggests that their evolutionary origins need to be sought in an earlier period. There is some evidence for this from burrows, thought to be those of a millipede, which have been found in fossilized soil dating from the late Ordovician. Because of these animals' closeness to forms now living one can be confident of their feeding habits. Spring tails (Collembola), bristle tails (Thysanura), mites and millipedes would have been scavengers, feeding on dead and decaying material and on fungi living on the detritus. Centipedes, spiders and the spider-like trigonotarbids were predators. (The latter group is extinct; they had massive jaws, and unlike spiders they did not produce silk.) None of them were plant feeders and at this time vegetation on land was sparse. However, seaweeds, filamentous algae, and animals flourished in the seas, lakes and rivers; tides, storms and

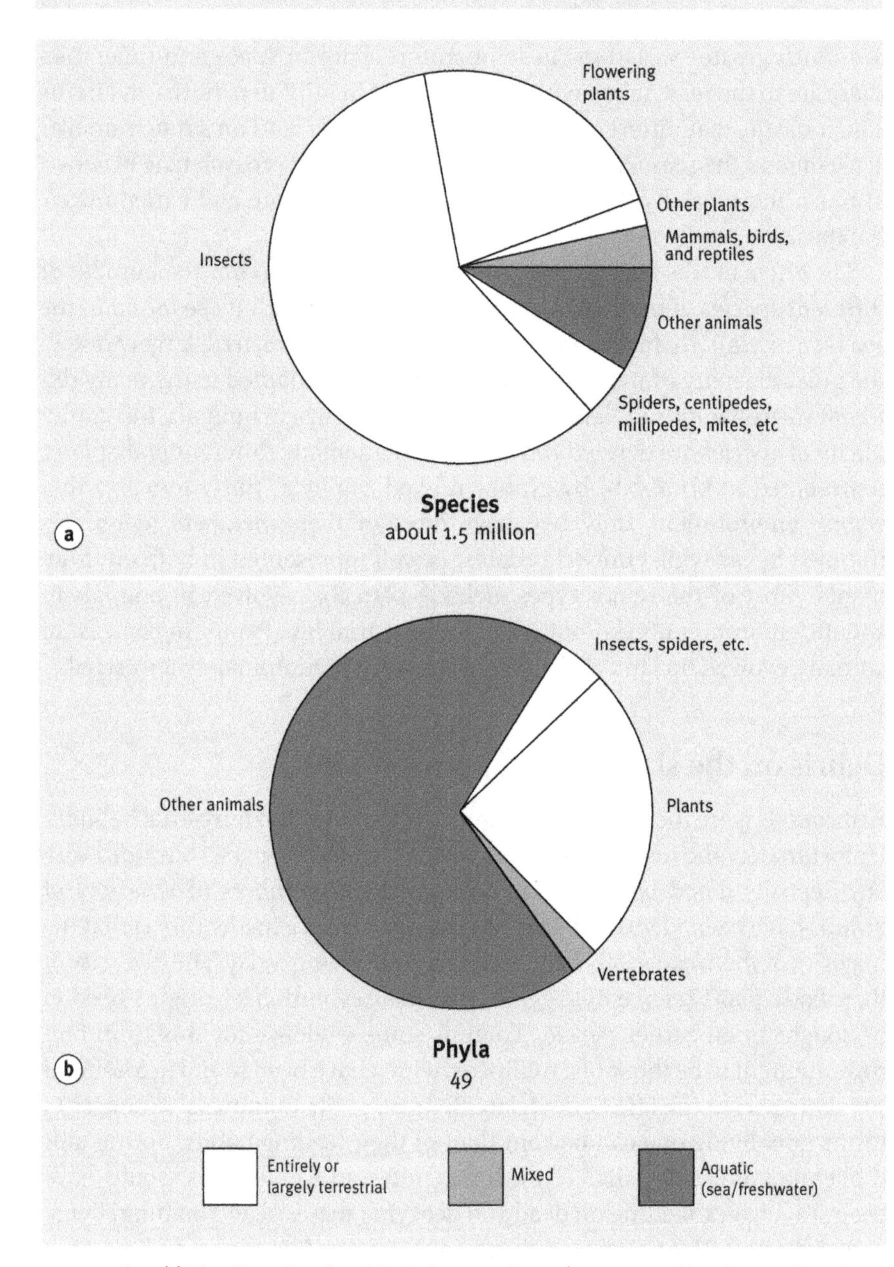

FIG. 6.7 (a) The diversity of multicellular organisms (represented by the number of different species) and (b) the disparity (shown by the number of fundamentally different body plans represented by different phyla)of animals and plants that are predominantly marine or terrestrial; ((a) modified from Southwood, 1978; (b) based on phyla as recognized in Margulis and Schwartz,1998)

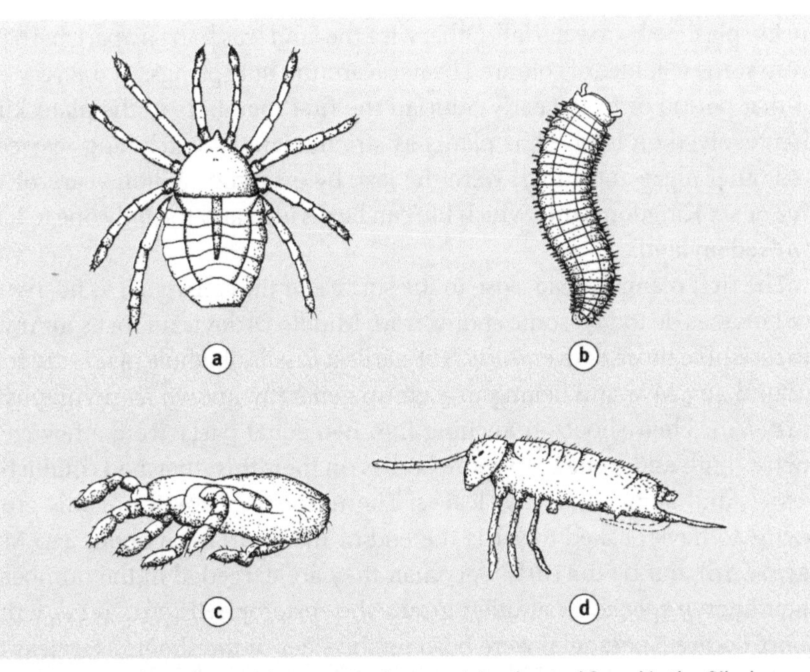

FIG. 6.8 Some members of the groups of terrestrial arthropod found in the Silurian and Devonian: (a) trigonotarbid; (b) millipede (Carboniferous); (c) mite (Devonian); (d) springtail (Collembola) (Recent)
((a) after Dunlop, 1996; (b) after Shear, 1997; (c) after Hirst, 1920)

floods would have left piles of debris on the shores. These piles would provide just the sort of humid conditions that would have been necessary for animals with recent aquatic ancestors and today members of these groups are abundant in and under such piles of debris. Indeed many extant species of these arthropods are not very resistant to water loss; most spring tails die within a few hours if exposed on a dry day with a relative humidity as high as 50 per cent. Thus the first terrestrial ecosystem would have been based on an input of detritus from aqueous habitats, the primary consumers were scavengers and these provided food for a few predators. The food chain was short.

From slime to trees

It seems probable that in the Ordovician the damp banks of rivers, coastal mud flats and similar habitats were colonized by cyanobacteria and simple algae, giving a thin covering of green slime such as one might find today on rocks splashed by a waterfall. Otherwise the land was barren apart, perhaps, from some colonies of coloured bacteria around hot springs. At the very end of that period or in the early Silurian the first members of the plant kingdom evolved on land. Thus plants as strictly defined (excluding seaweeds and other algae and fungi) were the last, by over 500 million years, of the five or six Kingdoms into which life can be divided and the only one to have evolved on land.

The first plants spread close to the surface of the soil, much as liverworts and mosses do today; some spores from Middle Ordovician rocks in Arabia do resemble those of liverworts. The earliest fossils of whole plants are from around 425 Mya and belong to a group generally known as rhyniophytes (FIG. 6.9). Their shoots, branching into two equal parts, were a few centimetres high and some bore spore bodies on their tips; they had conducting vessels and stomata, but no leaves. The number of species of this group seems to have peaked towards the end of the Silurian, at about 410 Mya. (FIG. 6.10), but by the early Devonian they are exceeded in the number of their known species by another group, the zosterophylls (FIG. 6.11), whose spore bodies (sporangia) were born on the sides of the shoots as well as the tips. When these shed their spores they did so along definite split lines and, in some, the two valves were of different shapes. Zosterophylls were at their most diverse around 398 Mya, after which they were overtaken in diversity by plants grouped as trimerophytes (FIG. 6.12). The stems of these plants contained far more water conducting vessels (xylem) than those of the earlier groups; they grew to greater heights, some up to nearly three metres. A most important feature, being a key step for the evolution of higher plants, is the development in this group of a main shoot with lateral branches, some of which bore terminal fruiting bodies. In the Middle Devonian this mixture of plants colonized mud flats and similar locations, which would have

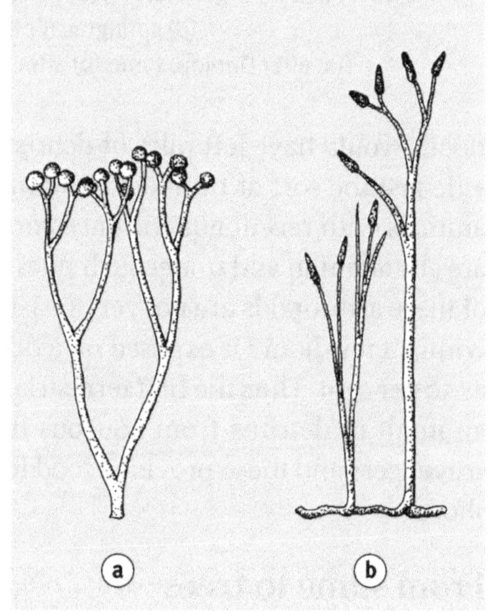

FIG. 6.9 Early plants (rhyniophytes) (a) *Cooksonia*; (b) *Aglaophyton* (after Cowen, 1995)

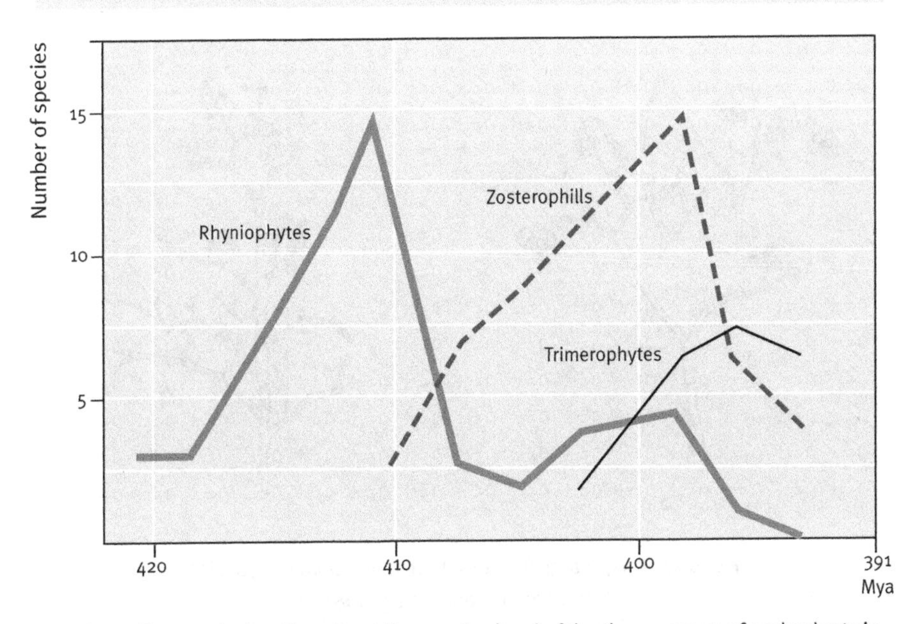

FIG. 6.10 Changes in the diversity at the species level of the three groups of early plants in the late Silurian and early Devonian (after Edwards and Davies, 1990)

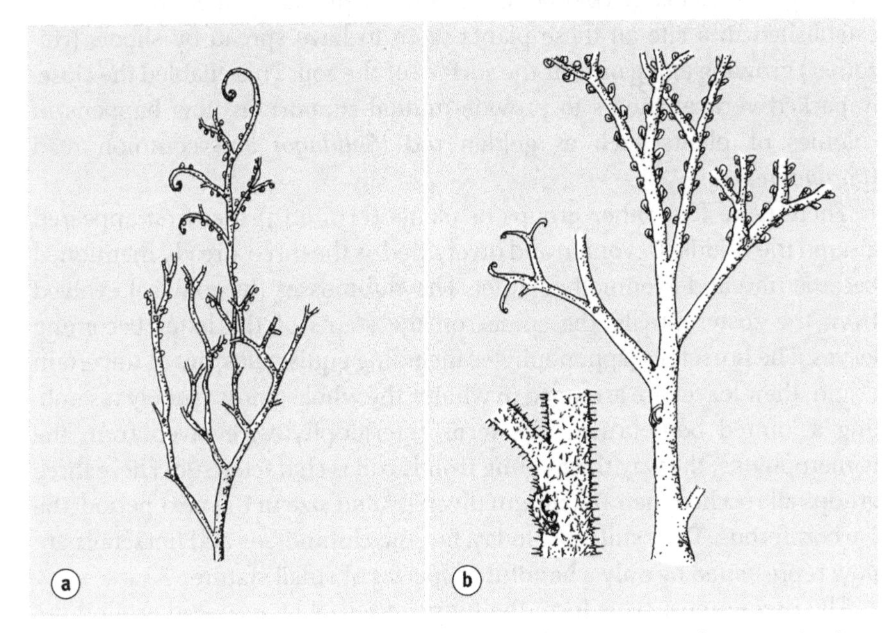

FIG. 6.11 Early plants (zosterophylls) (a) *Gosslingia*; (b) *Deheubarthia* (spines omitted on shoot, insert part of shoot showing spines) (after Edwards and Davies, 1990)

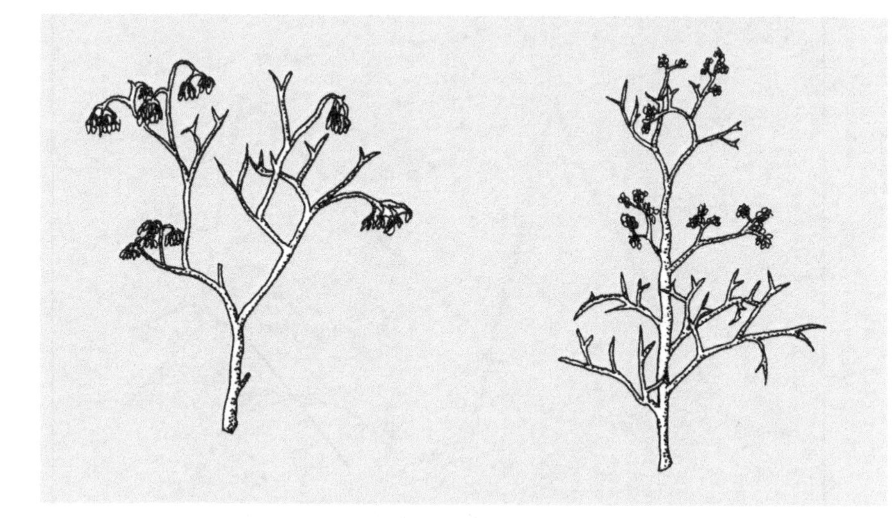

FIG. 6.12 Early plants (trimerophytes), *Psilophyton* species (after Thomas and Spicer, 1987)

had a meadow-like appearance with patches of the taller trimerophytes growing amongst large colonies of the shorter plants of other groups. Once established in a site all these plants seem to have spread by shoots (rhizomes) growing along or near the surface of the soil. This enabled the closely packed vertical shoots to provide mutual support, as now happens in colonies of plants such as golden rod (*Solidago*) and common reed (*Phragmites*).

There were four other groups of plants (FIG. 6.13) that first appeared around the Middle Devonian and diversified as the three already mentioned became rare and eventually extinct. The clubmosses (lycophytes) evolved from the zosterophylls, the spines on the stems of the latter becoming leaves. The horsetails (sphenophytes including equisetales) are of uncertain origin; their leaves are arranged in whorls, the whole plants crudely resembling a jointed bottlebrush. The ferns (pteridophytes) evolved from the trimerophytes; the way their young fronds coil is characteristic. These three groups all reached their maximum diversity and size in the next period, the Carboniferous. They still exist today, but the clubmosses and horsetails are now represented by only a handful of species of small stature.

The progymnosperms form the fourth group that appeared around the Middle Devonian, they evolved from trimerophytes. Unlike their contemporaries they became extinct within about 50 million years but they are of

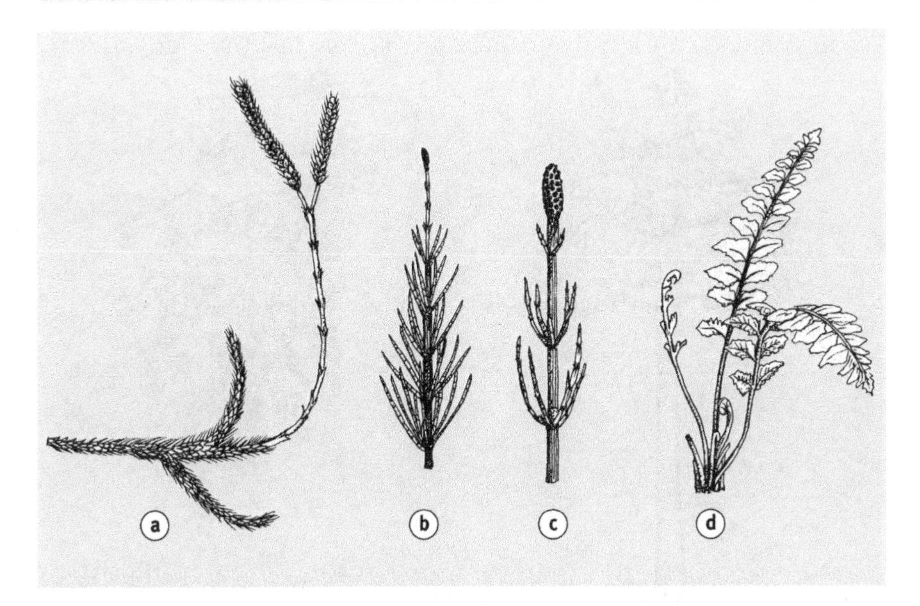

FIG. 6.13 Members of ancient groups of plants: (a) clubmoss (lycophyte); (b) & (c) horsetail (equisetales); (d) fern (pteridophyte)

great significance because they were the first trees and provided the stock from which all seed plants evolved. Just as one can argue that a certain group of the lobe-finned fish (p. 73) still live on as tetrapods or some dinosaurs live on as birds, so in terms of descent the progymnosperms live on in virtually all the plants we see today. *Archaeopteris* (FIG. 6.14), found in the fossil record from about 375 Mya, reached heights of around twenty metres. It is notable that once plants had evolved and established a toe-hold on land, within a mere (in geological terms!) 50 or so million years they had evolved from tiny straggling green shoots to majestic trees. This is a reflection of the evolutionary pressure to remain in the zone of maximum light. Plants compete directly and, even just within the Devonian period, a wave of one group of plants was replaced by another, and this occurred several times (as can be seen in FIG. 6.10). This leads to the conclusion that in plants one group may displace another by direct competition for space, the newcomer having evolved a better design; whereas in animals, in the majority of instances, it seems that a group had already died out or was in decline before it was replaced by another.

The key adaptation in trees was the evolution of the ability to add strengthening and conducting vessels to the ring of vessels in the stem, so

FIG. 6.14 One of the first trees, *Archaeopteris*, a progymnosperm (after Beck,1970)

creating a strong cylinder to hold the tree up, as well as a means of distributing water from the roots and food from the leaves. In the primitive plants the vessels were in the centre of the stem (protostele) (FIG. 6.15) as they are in the roots of all plants. This system has strength under tension—it resists pulling, but it has little rigidity. If you examine the roots of a plant you can see that they are much more flexible than the stem, which will have about the same amount of strengthening vessels but arranged in a cylinder (siphonostele). If you are unconvinced, take a flimsy sheet of paper and roll it into a tube. You can stand it upright and support a weight on the top.

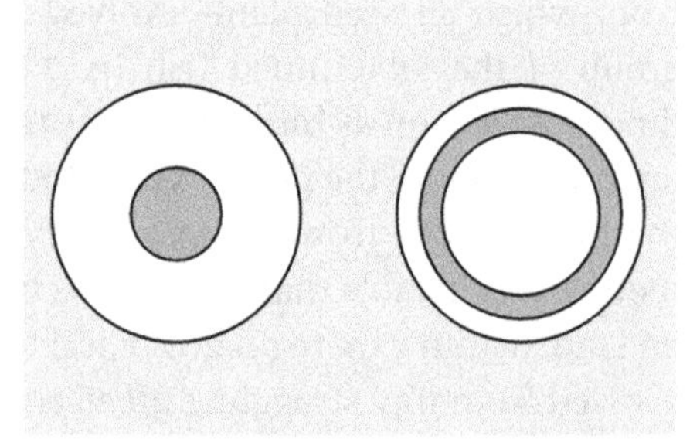

FIG. 6.15 Cross-sections of plants to show the different arrangements of the vascular bundles and strengthening tissues (dark tint): (a) protostele, as in the stems of early plants and in roots; (b) siphonostele, the basic type in the stems of most plants

The Devonian extinction

A large number of animal groups became extinct, or much reduced in diversity, towards the end of the Devonian period. Terrestrial organisms, particularly plants, seem to have been largely unaffected. The greatest impact was on reefs, certainly on two, but possibly on three, occasions over the last ten million years of the period. There were very substantial reefs off the coasts of what are now western Australia and western Canada. Their main builders, the stromatoporid sponges, became extinct and the reef-dwelling corals, both rugose and tabulate types, were also much affected. The armoured placoderm fish also became extinct at the end of the period, and the nautiloids were greatly reduced, while the conodont fossils show periods when most species became extinct followed by periods of rapid radiation, as new species occupied the new or newly available niches. Some other groups of marine animal, such as bivalve molluscs, were apparently little affected. Still further groups, such as the ancient jawless fish and graptolites, had declined throughout the Devonian so that it is not clear whether these events at the end of the period were the *coup de grâce*, or whether the animals had already vanished.

What was the cause of these episodes of extinction? An asteroid or other extraterrestrial body may have collided with the earth, but until the Woodleigh crater, near Shark Bay, western Australia, was discovered in 2001 there is no strong evidence. More research will be needed before we can be sure that this was the cause of a global extinction. It can be argued that a modest asteroid could have caused some of the local episodes of extinction, but anything global would surely have affected the plants.

Another suggestion is that, as—by the end of the period—tectonic movements had brought what is now the west of South America to lie on the South Pole, the resulting ice sheet formed would have led to global cooling and a lowering of sea level. While this was undoubtedly a major component of the general environment at the time, it is difficult to regard this continuing process as a sufficient explanation, because there seems to have been something of a 'recovery' between the first and second extinction events. Also the two episodes had different effects on different groups: for example the deep water rugose corals were much less affected by the first episode than the second. Tectonic movements had another outcome. Gondwana and the main part of Laurasia had just started to collide, and the remaining components of Laurasia were moving together. This must have had major effects on the flow of the warm and cold ocean currents; changes could occur relatively rapidly. In the late Devonian there were many straits between the different land masses and one can envisage that over a short period of geological time a coastal current could change from warm to cold and back again. There was also the vast Panthelassic Ocean and upheavals in its bed could have had major effects on sea level. When sea level rises relatively abruptly and anoxic water from the deep ocean is carried into the well-oxygenated coastal shallows, the increased level of sulphur compounds causes the deposits to become black. Some of the Devonian reefs are overlain by black deposits; but it is not known whether this was a result of a fairly rapid rise in sea level, so that the corals were virtually suffocated, or whether this happened after the corals had already died from a fall in temperature and a subsequent rise in sea level brought anoxic conditions.

It seems most likely that the series of extinctions resulted from a variety of changes brought about by the tectonic movements that led to the assembly of the giant continent Pangaea. What is particularly interesting is that, although of course tectonic movements continued, over a hundred million years were to pass before another mass extinction. This occurred when the giant continent was finally assembled and this was the greatest extinction of all.

CHAPTER 7

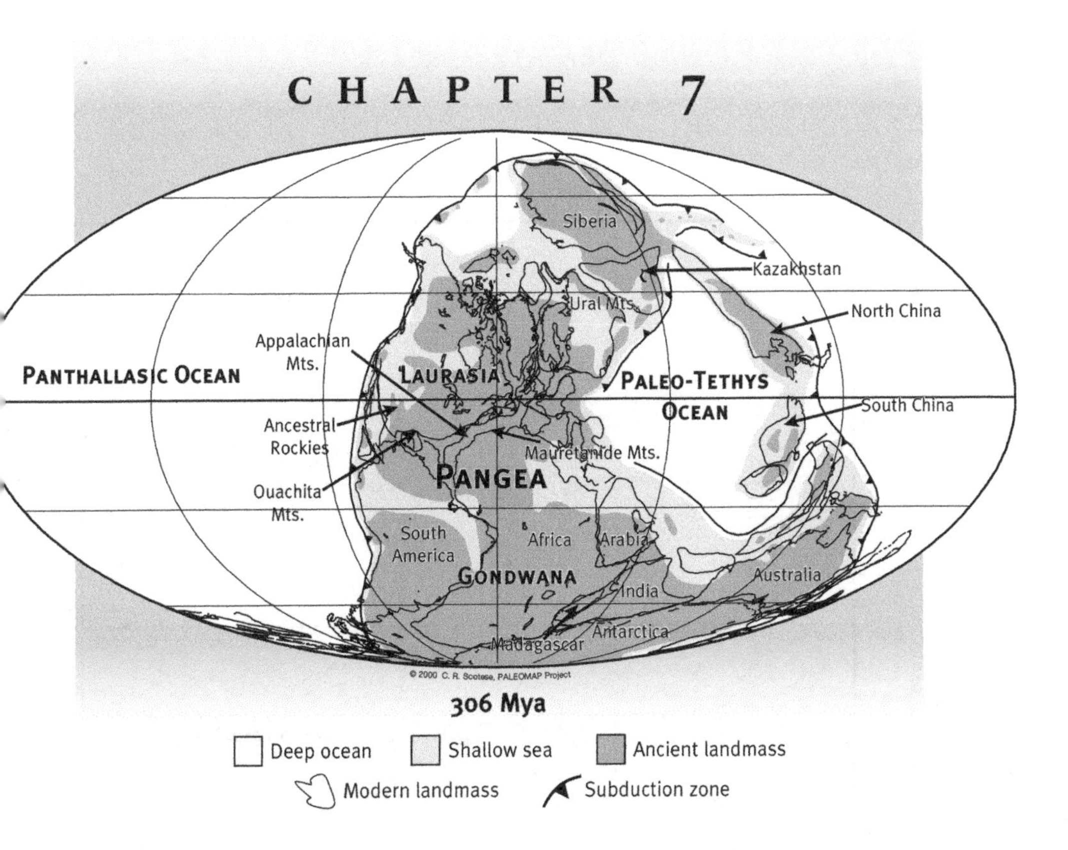

The Giant Continent Forms

Carboniferous–Permian
362–248 Mya

AT THE start of this period Gondwana and Laurasia had just started to collide; by the end of it, the giant continent of Pangaea stretched from the South Pole to near the North Pole with the Palaeo-Tethys Sea on its eastern side, almost entirely enclosed by 'continents' that have come to be parts of China and south-east Asia. In the tropical regions, around the Palaeo-Tethys Sea there were, particularly in the Carboniferous period, extensive forested swamps, now represented by deposits of coal, oil and gas; later in the Permian period coal was deposited further away from the equator. In the early Carboniferous an ice sheet formed over the southern portion of

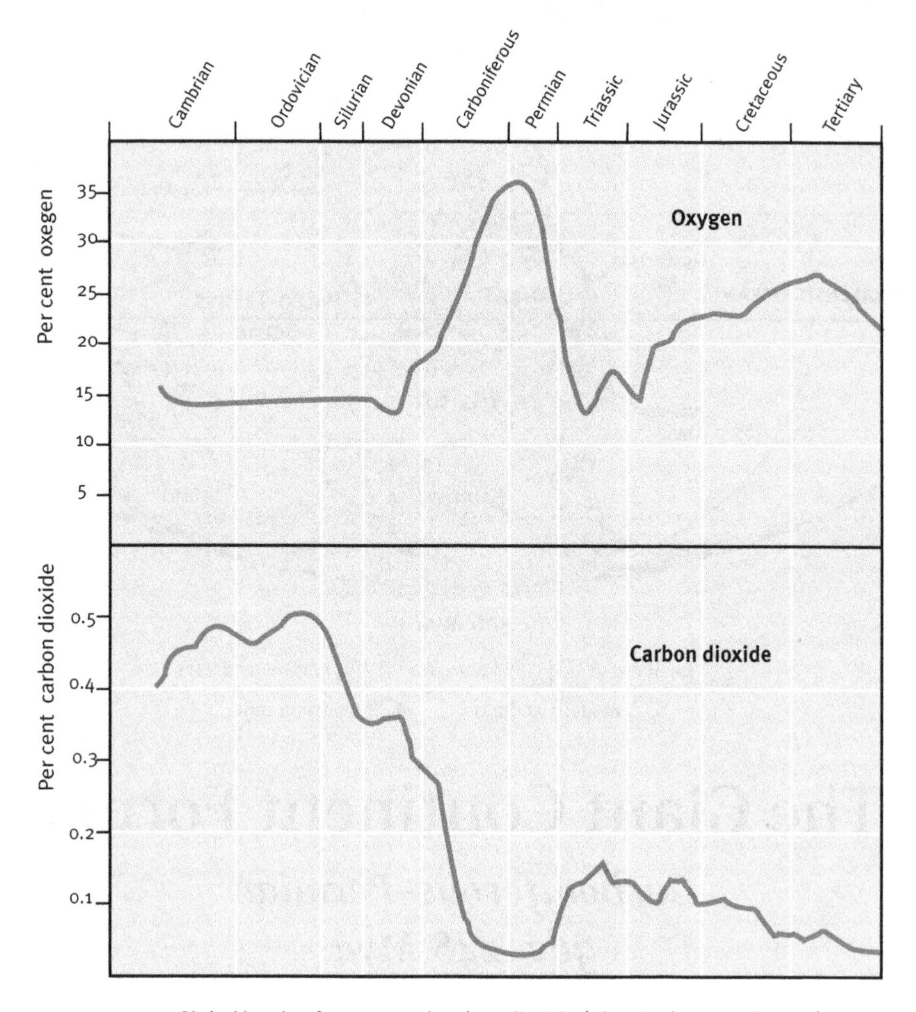

FIG. 7.1 Global levels of oxygen and carbon dioxide (after Graham *et al.*, 1995)

Gondwanaland which became increasingly more extensive until the late Permian. The extent of any ice sheet over the northern extremity of Pangaea, that was within the arctic circle and is now Siberia, is uncertain. By the end of the Permian, sea levels were low, partly due to the compressive forces acting on the supercontinent and partly because of the water held in the polar ice sheets. But throughout the period, particularly in the late Carboniferous, there were repeated, and little understood, episodes in which the sea level fluctuated—the forests forming and then being drowned when the level rose.

Perhaps the most unique physical features of this period were the rise in the amount of oxygen in the air and the associated diminution in the proportion of carbon dioxide (FIG. 7.1). These are considered to be due mainly to the large quantity of dead plant material turning into peat which, after heating and compression, has formed coal deposits. This material is the product of photosynthesis by which process carbon dioxide is taken in and oxygen given out. During most of the earth's history when plants have died they have been oxidized in the process of decay: oxygen is used up and carbon dioxide given out. The complete process is termed the carbon cycle (FIG. 12.9), but during this exceptional period the bulk of the carbon was 'banked' as peat so leaving the oxygen free in the atmosphere. The reduction of the carbon dioxide level would have led to a 'reverse greenhouse effect', with more heat being reflected back from earth into space. As a consequence of this and of the relatively large amount of land in the polar regions, global temperatures were relatively cool for most of the time covered in this chapter.

Great forests

The first trees evolved at the end of the Devonian period, but this is truly the age of forests. Their development had profound implications for other forms of life, contributing to or even leading to, the evolution of flight and of bipedalism (the ability to stand on the hind legs). But their key effect is that they greatly increase the amount of living space in an area, because like coral reefs, they provide ecospace. The magnitude of this cannot be precisely measured on fossil trees, but it is shown by some measurements made recently on an old, but not particularly large, beech tree. The tree's trunk occupied six square metres of soil surface, but the total surface area of the tree was calculated as eleven thousand square metres; an eighteen-hundred-fold increase in space. This of course included the upper and lower surfaces of the leaves, the twigs and the bark, and the surface area under the bark on the dead branches. Other organisms will have evolved to live on these surfaces. All the surfaces may be colonized by fungi of one sort or another. Plants, like ferns and mosses, will grow on the trunk and large branches, they are termed epiphytes. Small animals will be found in all these places, different species will have different niches. There are, for example paper-thin bugs that live under the bark on dead branches and feed on the fungal filaments growing on the wood. Larger animals like birds and squirrels will be limited to the larger structures—branches and the trunk—but even for them a tree provides much more living space than the ground it occupies. The ground shaded by a tree's branches is also still available to shade-tolerant plants and to soil and ground-dwelling organisms, ranging from fungi to deer. An old tree may have holes in its branches, perhaps a hollow trunk, which provide important refuges for many vertebrates, such as snakes, birds, bats and rodents; many different species may make their home in a single old trunk in a tropical rain forest. These associations were quickly established once trees had evolved as is evidenced by the lizard-like reptiles found fossilized in hollow tree trunks from the Carboniferous period.

The arborescent (tree) form poses certain biomechanical challenges that are best considered in relation to trees as a whole, not simply in respect of the earliest trees. Basically, the problem is that of holding all the weight above the ground. Firstly, there are the large forces induced by the tree's self-weight. As has been previously discussed (p. 86), by having its strengthening tissues in a cylinder the tree gets maximum support. Applied loads such as rain and snow also increase the burden on the tree. In regions where there is heavy rainfall, tree leaves are commonly smooth, downward pointing with a drip tip (see FIG. 10.2) so that surface water is quickly shed. Trees like the spruce, that grow in snowy places have branches that bend down easily, thus shedding much of the snow that falls on them, rather like the steep roofs of a chalet.

Wind provides a second challenge. A tall cylinder exposed to a force from the side will come under tension. If the force is increased it will eventually snap on the side where the force is applied and buckle on the other. But extra strength can be given by reinforcing with another material to take the tension. In tree trunks and branches this strengthening is provided by long thin cells (vessels) that have thickenings in their walls (FIG. 7.2(d)). During development it seems that these thickenings, especially the spiral ones, are 'seeking' to expand, but are held by surrounding tissue; in this way the wood is prestressed and can withstand tension (FIG. 7.2(a)–(c)). It is the outer tissues of the sap wood that provide this prestressing. This characteristic is known to many who work with wood; the old bows found on the Tudor ship, the 'Mary Rose', had the sap wood on the convex side of the bow so giving the maximum strength in tension. If the surface of bark and sapwood is scraped off a young twig of elder, or many other trees, it curls up. This is because now that the sapwood can expand, it is no longer held by the

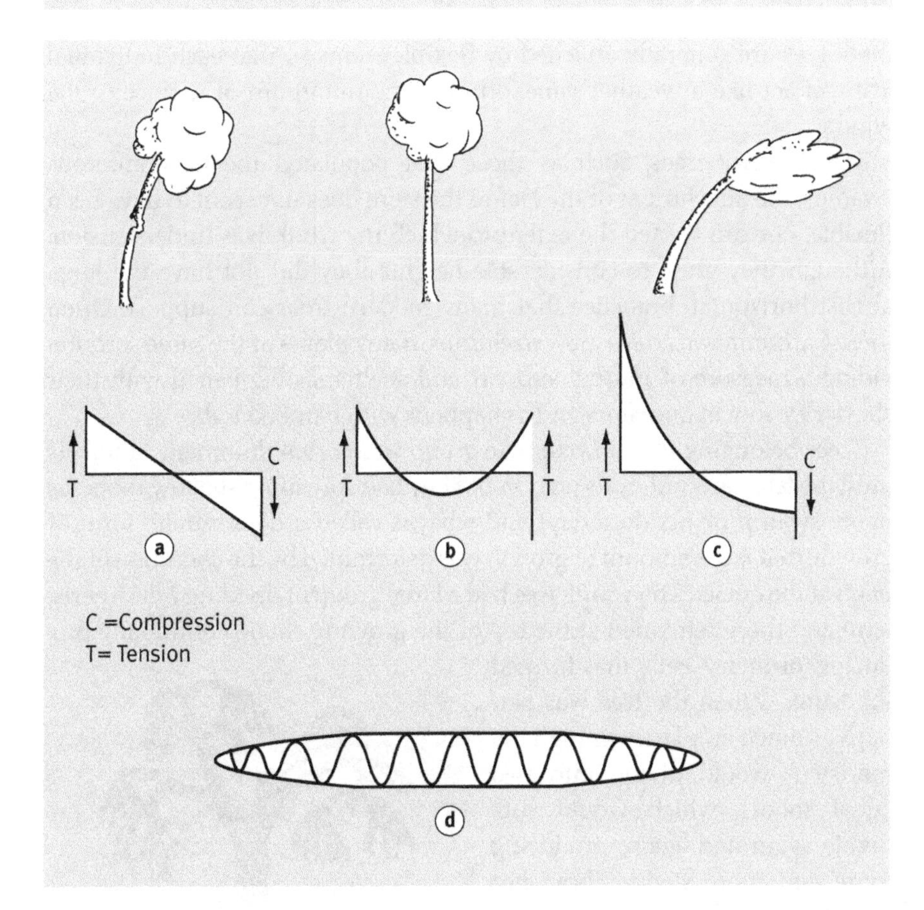

FIG. 7.2 Mechanics of a tree; (a) hypothetical tree in a gale without any of the adaptations that have evolved, in particular no pre-stressing in the wood; (b) a living tree in calm with pre-stressed trunk; (c) a living tree in a stronger gale than 'a', the pre-stressing reduces the compression stress and elevates the snap point, the branches and leaves also turn to minimize their wind resistance; (d) a spiral vessel ((a)–(c) modified from Gordon, 1978)

wood on the inside, but it is still constrained on the other side by the bark which has a fixed length.

In engineering terms, a tree is a singly connected structure, therefore any force applied by the wind is reduced: each part of the tree may turn and bend as much as it needs without distorting or constraining another part. If we look at a tree in the wind we can see that all the branches and twigs tend to turn away from the wind so minimizing the resistance. On modern trees the leaves are generally attached by flexible stems so that each individual leaf can act like a weather vane, offering the minimum of surface to the wind.

In the early trees, such as those that populated the Carboniferous swamps, the attachment of the leaf to the stem does not seem to have been flexible, nor can we tell the extent to which the trunk was under tension. Although they grew to considerable heights they did not have the long, almost horizontal, branches that many modern trees can support. Often these Carboniferous trees grew in clumps, many plants of the same sort providing a measure of mutual support and shelter, as happened with their shorter Devonian ancestors and as happens with bamboo today.

Trees belonging to the Lycophyte group were often dominant in forests until the late Carboniferous (FIG. 7.3). They had spreading shallow roots (as many swamp plants do today) and what is called a determinate form of growth, that is the amount of growth was determined by the condition of the plant at the outset. The young tree had a large group of dividing cells (meristem) and these remained at the top of the growing shoot, continually producing 'ordinary' cells that formed the trunk. When the tree was perhaps as much as 40 metres high the meristem would divide into two equal shoots, which would subdivide again and again, producing more and more slender branches until the meristem could no longer be divided. The tree effectively ran out of cells that had the potential to divide. One can see that this system lacked the adaptability of modern trees whose shoots have indeterminate growth: in these there are two sorts of meristem, primary and secondary. A small part of the primary meristem remains at the end of every shoot. The increase in diameter of the twig or trunk is secondary, in that it arises, not from the primary meristem but from other

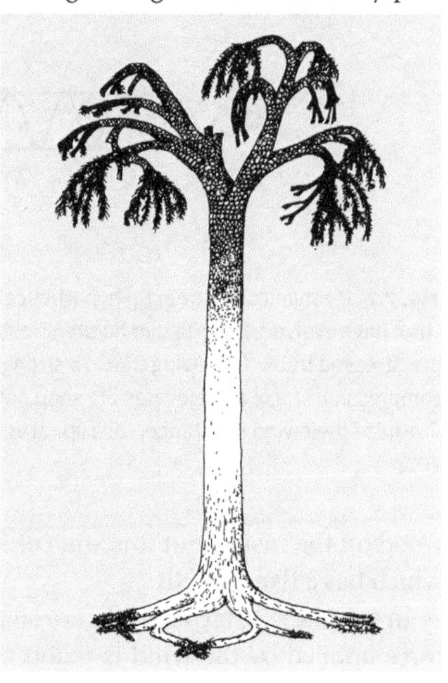

FIG. 7.3 Tree with a determinate growth pattern, *Lepidodendron* (after Thomas and Spicer, 1987)

cells (the secondary meristem) that remain throughout the shoot and retain the ability to divide.

The horsetails (Sphenophytes) were represented in these forests by a number of giants, such as *Calamites* (FIG. 7.4) which could be up to thirty metres tall. Like modern horsetails, they grew from underground shoots (rhizomes) and formed clumps. The network of rhizomes would serve to anchor the clump in the soft mud. These larger horsetails did not survive into the Permian, possibly because they were not suited to the drier and more compact soils. Other, smaller horsetails did and some have been found in sites where conditions would have been temperate, possibly even polar.

Ferns evolved at the end of the Devonian and have remained an important element in the flora of many regions ever since. Their light spores are carried great distances in the wind and although often associated with damp places, ferns can withstand extremely dry conditions. Bracken is one of the most widely distributed plants today. Tree ferns up to about eight metres in height were frequent in Carboniferous forests and the group was represented by many other different types: the period has also been called the 'age of ferns'.

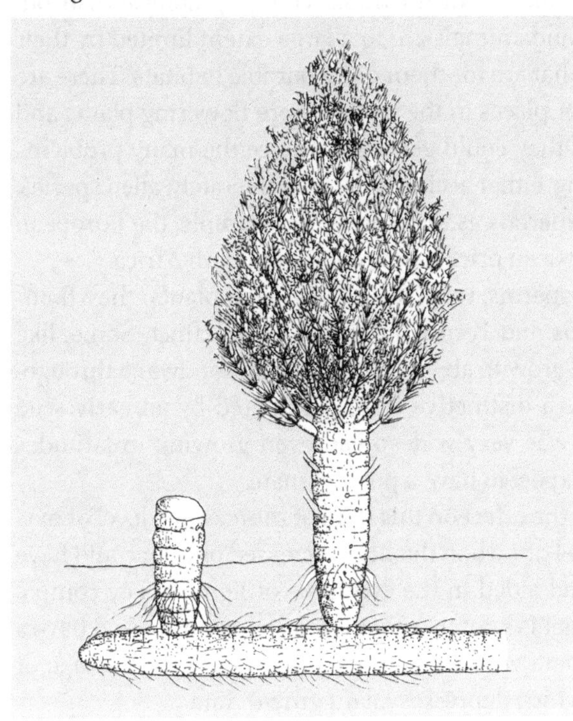

FIG. 7.4 Giant horsetail, *Calamites*

The three groups (lycophytes, horsetails and ferns) already described are spore plants, producing spores that germinate and grow into what is normally a very small flat plant (prothallus) on which sexual reproduction occurs, the male cells requiring a thin film of water through which to swim. This restricts them to habitats that, although they may be very dry for much of the time are, at least sometimes, moist. Some progymnosperms produced two sizes of spore. To simplify the story, the small spores eventually evolved to become pollen and the large ones (megaspores) the ovule of the flower and, often with the addition of surrounding structures, the seed. The development of the seed was undoubtedly a major step in plant evolution. Although seeds are heavier than spores and so are not transported as far on the wind, they can withstand bad conditions and contain a reservoir of food that aids the establishment of the young plant. Modern plants have, of course, a wide range of evolved mechanisms to ensure their dispersal, but floras dominated by seed plants tend to be more local than those of ferns, and became characteristic of a particular locality; for example, in the northern hemisphere, evergreen oaks grow in areas with a Mediterranean climate, while eucalyptus flourish in Australia. The geographical distributions of terrestrial plants and animals are to a large extent limited by their inability to travel across what are for them unfavourable habitats. There are therefore often many other places in the world where flowering plants and animals would flourish if they could get there. Hence the many problems created by man introducing, either accidentally or deliberately, alien species. They often become so numerous as to be pests, for example, the European rabbit in Australia, the Mexican prickly-pear cactus in South Africa.

The seed ferns, pteridosperms, were the earliest seed plants; they flourished in the Carboniferous and Permian, but are now extinct. Some, like *Medullosa* (FIG. 7.5) could grow to about ten metres. In Gondwana throughout this period, there was a distinctive flora dominated by an early seed plant *Glossopteris* which was very widespread even growing in latitudes (80°–85°) that one would expect to have a polar climate.

It is difficult to surmise the effect on this flora of the elevated level of oxygen and the depressed level of carbon dioxide. Increased oxygen could have diffused into the trunk and aided in the synthesis of lignin, a key component of woody tissues. What is known is that leaves from this period have a very high density of stomata which would have facilitated the passage of carbon dioxide, a lack of which depresses plant growth rate.

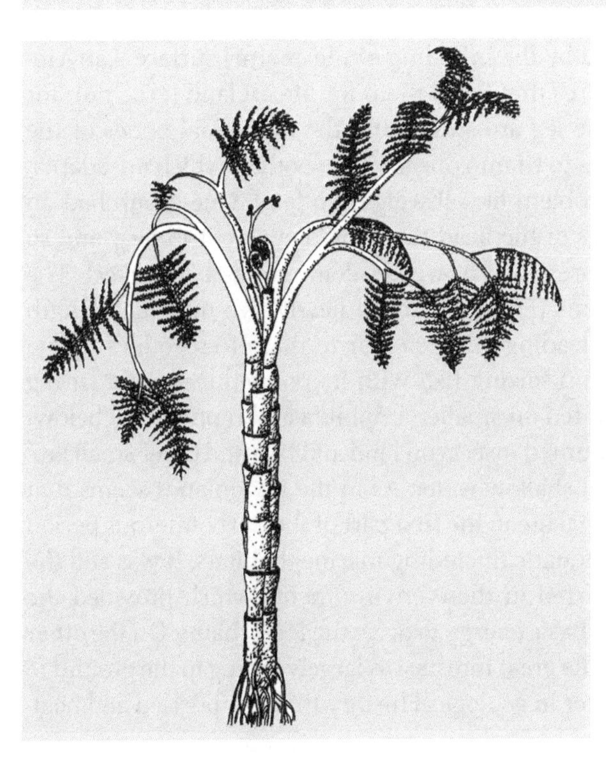

FIG. 7.5 Seed-fern, *Medullosa* (after Thomas and Spicer, 1987)

Amphibious life

The amphibious lifestyle can be defined as that of an individual spending its life partly in an aquatic environment and partly in the air on dry land. Two groups with this lifestyle became fully established during this period. Firstly, there were those four-legged vertebrates (tetrapods) that are conveniently grouped together as amphibians and, secondly, certain insects, particularly those related to dragonflies and mayflies. It is also likely that eurypterid sea scorpions, that then frequented freshwater more than the sea, may have hauled themselves out on muddy banks.

We noted in the last chapter that those Devonian animals with feet rather than fins, and the lobefinned fish closest to them, probably lived in lagoons, tidal flats and deltas and used their limbs to haul themselves through dense vegetation and occasionally took gulps of air, as fish may do today. After the Devonian there is a gap in the fossil record of over 20 million years, by which time a variety of amphibians are found. Most of them can be loosely grouped as 'labyrinthodonts', a name which describes the pattern of wavy lines in their teeth caused by the infolding of the enamel surface. Labyrinthodonts had many features that fitted them for life on land (FIG. 7.6): for example, the bones of the leg are strong and the individual bones of the backbone are jointed so as to fit into one another, both of which are adaptations to overcome the problem of self-weight on land. One group had an arrangement in the bones of the head that would enable it to hear and so speculations that these forests were largely silent may not be correct. The largest of these amphibians (*Eryops*) reached nearly two metres in length and one can visualize it leading a life similar to that of crocodiles today; resting in the shallows and seizing fish with its pointed teeth and strong jaws. It undoubtedly also fed on smaller amphibia and reptiles (see below) that in turn would have hunted insects on land and fish and other small animals (such as shrimps) in shallow water. As in the Devonian it seems that the food chain for terrestrial life in the first part of the Carboniferous period depended ultimately on aquatic (including marine) habitats. It was still the photosynthesis that occurred in these environments which provided the primary productivity, the basic energy to drive the food chains. On the other hand the productivity of the great forests was largely falling to the ground to form deposits of peat. Later in geological history this was pressed and heated and formed coal.

It is assumed that the labyrinthodonts, like modern amphibians, laid their eggs (with a gelatinous cover) in the water. This is where the tadpole stage was spent. Having gills they would have obtained their oxygen from

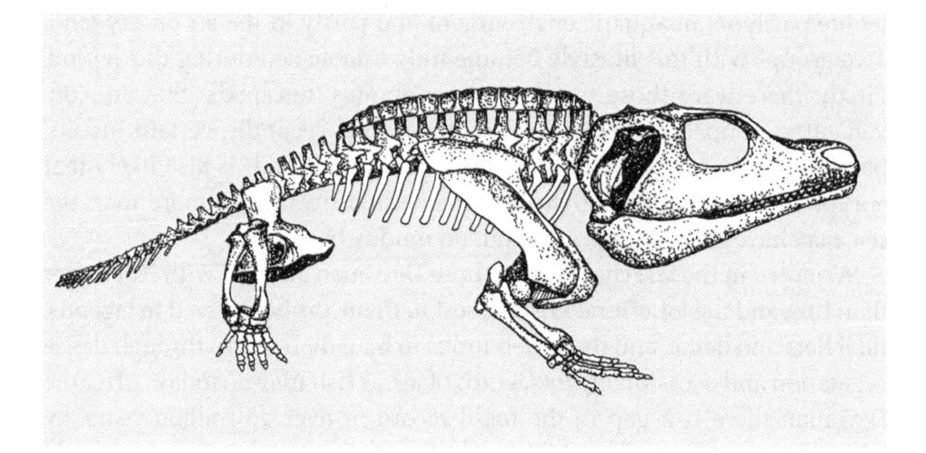

FIG. 7.6 Amphibious tetrapod, *Cacops*, (after Cowen, 1995)

the water. In spite of this dependence on water, the labyrinthodonts lasted through the arid Triassic period and one (*Koolasuchus*) has been found in Australia in the Lower Cretaceous. Modern amphibian groups are not known from the fossil record until the late Triassic or Jurassic, though they must have evolved earlier from a different group of amphibians found in the Carboniferous.

The high level of oxygen in the air at this time would have aided the functioning of early lungs and facilitated some breathing through the damp skin—as occurs in modern frogs.

The mayflies and dragonflies are two of the groups of winged insects that lay their eggs in water, where their larval stages live. They have a complex wing venation that indicates that they are the most primitive of winged insects; fossils of mayflies and dragonfly-like insects have been found in Carboniferous rocks (FIG. 7.7(a)). It seems likely that these insects were at one time entirely terrestrial, their young living, like the earliest insects found in the Devonian, in very damp conditions under debris. The direction

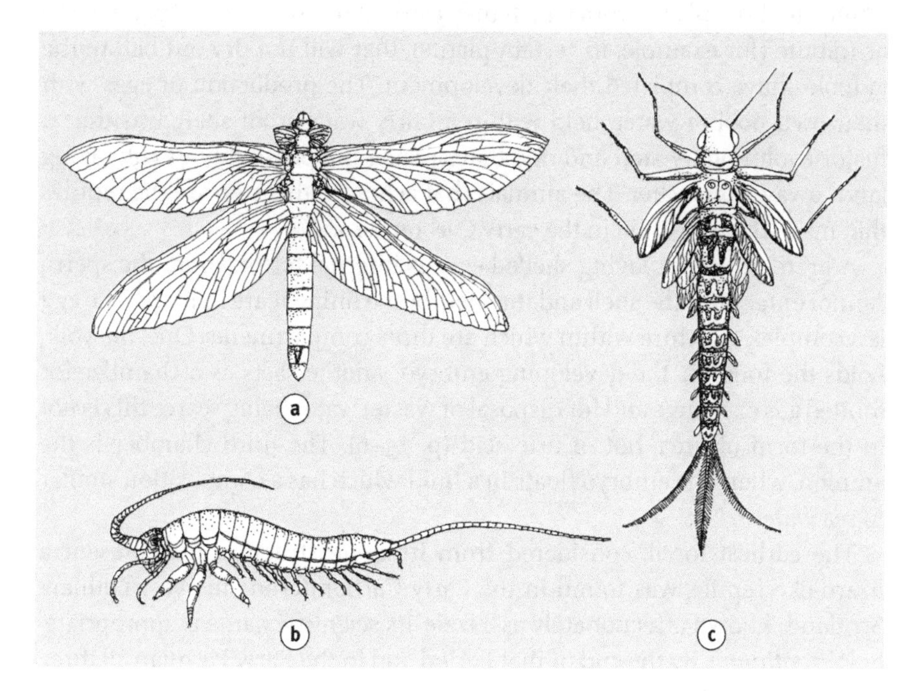

FIG. 7.7 Carboniferous and lower Permian insects: (a) dragonfly (an eugeropterid), antennae and legs not reconstructed; (b) bristletail (*Dasyleptus*); (c) mayfly larva (aquatic) (after Kukalova-Peck, 1985, 1987)

of evolution then reversed so far as the larval stages were concerned, for they returned to water—mayflies clearly had aquatic larvae in the Carbon-ifer-ous (FIG. 7.7(c)). These insects were giants: dragonfly-like species existed with wing spans up to 70 centimetres—the largest wing-span of any known insect, more than twice that of the largest of today's insect (a moth). Some other fossil insects from the Carboniferous are also large relative to the present size of members of their group; the largest of the Carboniferous mayflies had a wing span of 45 centimetres, about ten times that of modern forms; while the 35 millimetre long bristletail, *Dasyleptus* (FIG. 7.7(b)) was about twice the size of modern species. The existence of such large insects probably depended on the elevated level of oxygen in the air that would have allowed their long air pipes (trachaeae) to work efficiently.

Eggs in shells

Amphibia need free water in which to spawn, so their occupation of land is limited to those places from which they can still reach rivers and pools, large or minute (for example, in certain plants), that will not dry out before the tadpoles have completed their development. The production of eggs with their own pool of water, held within a fairly waterproof shell, was thus a major evolutionary step and one that opened the possibility of colonizing lands away from water. The animals that achieved this step are the reptiles that made the transition in the early Carboniferous.

A prerequisite for laying shelled eggs is internal fertilization; the sperm cannot enter once the shell and the membrane lining it are in place. The egg is a complex structure within which are three compartments. One, the yolk, holds the food for the developing embryo; another acts as a chamber for limited gas exchange and for disposal of waste; water being scarce this is not in the form of urea, but of uric acid (p. 75–6). The third chamber is the amnion, where the embryo floats in a fluid which has a composition similar to sea water.

The earliest fossil, considered from its bone structure to represent a lizard-like reptile, was found in the Early Carboniferous in West Lothian, Scotland; known affectionately as 'Lizzie' its scientific name is appropriately *Westlothiana*. By the end of that period and in the Early Permian all three major groups of reptile had evolved. These groups are recognized by features of the skull: the number of holes behind the eyes being zero, one or two. Those without holes (Anaspids) are now represented by the tor-

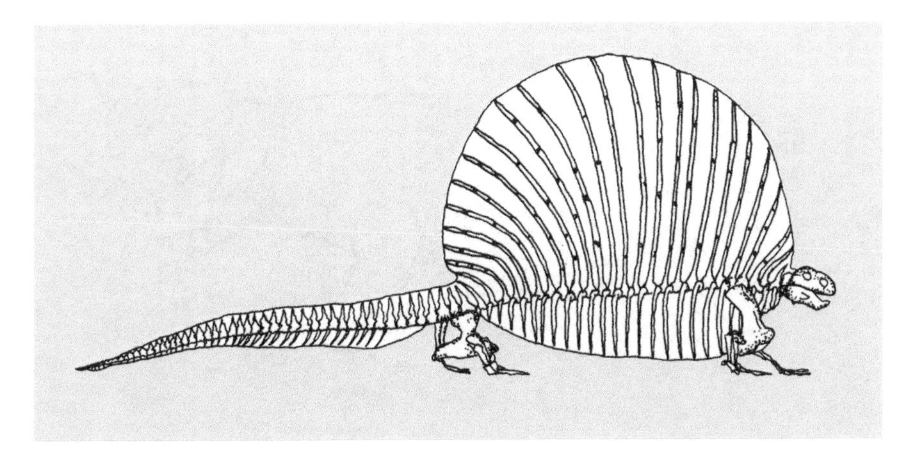

FIG. 7.8 Herbivorous pelycosaur, *Edaphosaurus* (after Cowen,1995)

toises and turtles; those with one (Synapsids), by the mammals and those with two (Diapsids)—having evolved into the now extinct dinosaurs and pterosaurs *en route*—by the birds.

In what is now North America the dominant reptiles of the Late Carboniferous and Permian were the 'sail-backs', the pelycosaurs (FIG. 7.8). The conspicuous 'sails', extensions of the vertebrae of the back, are believed to have functioned like solar panels, raising their blood temperature. This would enable them to 'speed up' to capture prey or escape enemies. Animals with little bulk can raise their temperature quickly by 'sunning' themselves as modern lizards and snakes do. The pelycosaurs however tended to be large, up to three metres in length so at times they may have also used their 'sails' like a car radiator—to cool down. Most of them were predators (i.e., carnivorous), but the different tooth structure of *Edaphosaurus* (FIG. 7.9(b)) identifies it as the first vertebrate herbivore.

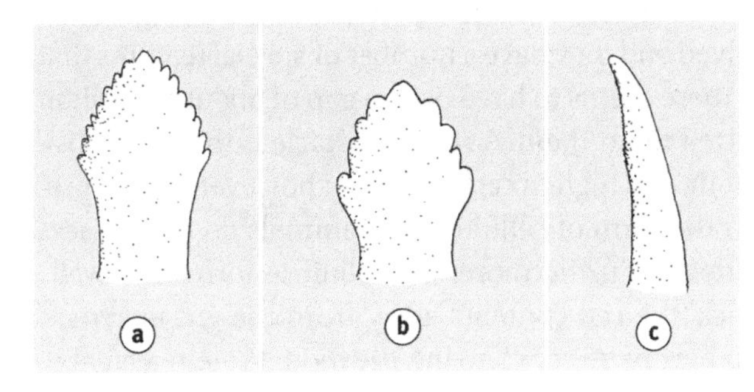

FIG. 7.9 Teeth: (a) modern herbivorous reptile, iguana; (b) *Edaphosaurus*, a herbivorous pelycosaur (c) carnivorous dinosaur, the inner curve is finely serrated like a steak knife

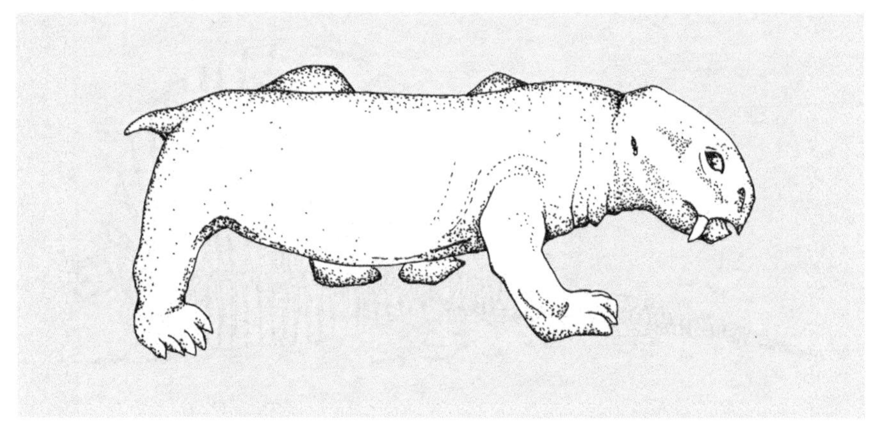

FIG. 7.10 Dicynodont (*Diictodon*) (after Cluver,1978)

During the Permian a different branch of the synapsids developed, which are known as the mammal-like reptiles or therapsids. One division of these, the dicynodonts, contained the most abundant herbivores of the Permian (FIG. 7.10). They were stocky animals, ranging in size from that of a rabbit (0.4 m) to that of a large rhinoceros (3.6 m), with a powerful head and two substantial tusks in the upper jaw. Whether these tusks were used for feeding—grubbing up roots, uprooting shrubs and trees—or whether they were used in aggressive combat or defence, is a subject of speculation. Thus, whereas until the Late Carboniferous there were no large herbivores to 'process' plant material, by the end of the Permian they were widespread. A typical aerobic food chain was established: plants, the primary producers, being eaten by herbivores themselves preyed upon by carnivores.

Eating plants—a new way of life

Plants, as we have seen, were the last of the Kingdoms into which organisms are classified to have evolved and they have a number of special features that account for the fact that there seems to have been a gap of about 50 million years before animals started to eat them. A major obstacle is their composition: they are low in protein, and high in carbohydrate; however, a large proportion of the latter is in the form of cellulose and animals do not possess the enzyme to break it down. Furthermore, the cellulose forms the walls of the cells and these 'lock' the cell contents away from the gut enzymes. Hence the importance to herbivores of chewing the food. Protein availability is often a limiting factor for herbivores. This is demonstrated by their faster development when their diet is supplemented with protein; hence the practice of feeding meat and bone meal to cattle that led to the spread of BSE ('mad cow disease') in the UK. If plants are fed abundant nitrogen fertiliser they have more protein in their tissues and the insects living on them often develop faster. This supplementation of the diet with additional protein also happens naturally; many animals (insects and vertebrates) that are predominantly herbivores also feed on other animals—living or dying. This omnivorous habit is found, for example, in mirid plant bugs, iguanas and pigs; it is particularly widespread in birds, where the diet of the young often contains a higher proportion of insects than that of the adult.

The obstacle of coping with cellulose and the shortage of protein has been overcome in most herbivorous insects and vertebrates by the evolution of a symbiotic relationship with a micro-organism. The gut often has special pouches or sections in which these organisms live (FIG. 7.11) and where they digest the plant material, providing a nutritious brew for their host. They may be likened to biochemical 'brokers' or 'middlemen' taking material from plants and passing it on, in a more useful form, to the animal. The development of such intricate symbiotic relationships must have been a major obstacle to the evolution of herbivory. Some insects prefer plant material that is diseased or rotting when its composition will already have been slightly altered by the micro-organisms and one can see this as a stage on the way to a symbiotic relationship. Herbivorous vertebrates may have gained their symbionts via insects, for the closest relations of the first herbivores were insectivores. Plant feeding developed first in insects and these insectivores would have taken in symbiotic micro-organisms when they fed. When there were vertebrate herbivores, other vertebrates could have become 'infected' through either predation or scavenging on dung (coprophagy). What is notable is that though there was a long gap between the evolution of plants and that of the first vertebrate herbivore, once one had evolved the habit quickly (in geological terms) developed in a variety of different groups.

Without symbionts there are two other ways that animals can get an adequate diet from plants. One is by selecting those parts of plants that are richer in protein—the growing shoots, the flower and the seeds and fruits. The other is to increase throughput which demands a high rate of feeding and the evolution of a large gut, as with caterpillars and geese. Even those with symbionts need a large gut to hold the food while the micro-organisms

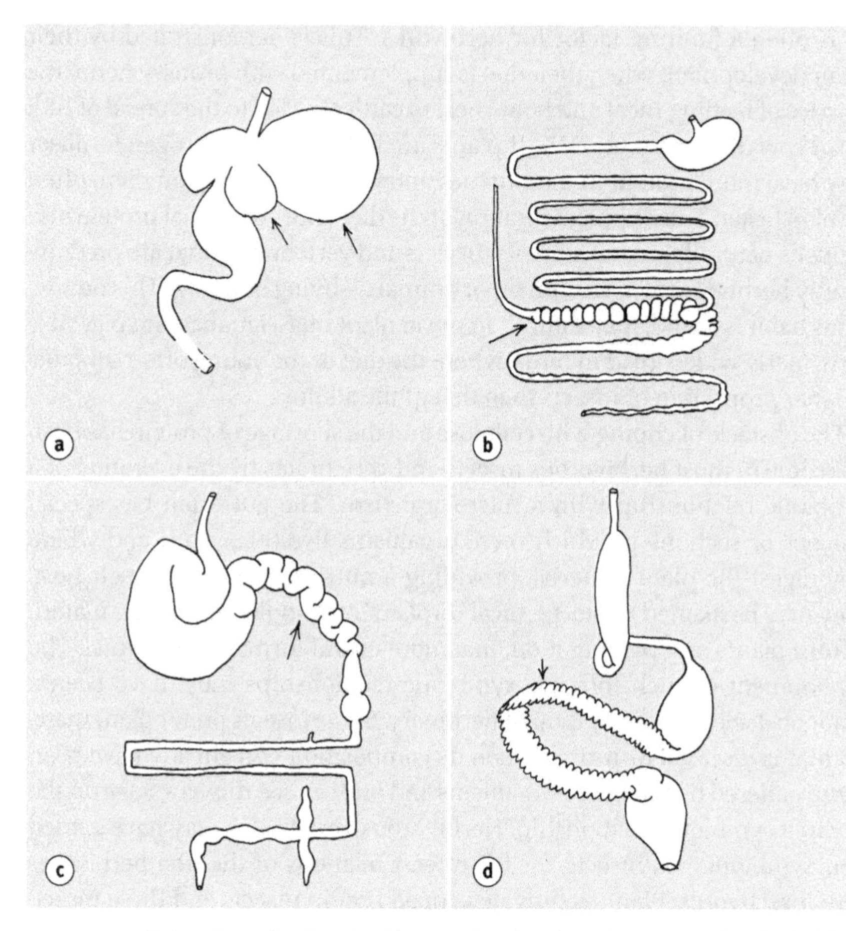

FIG. 7.11 Gut system of various herbivores, showing the enlargements (marked with arrows) occupied by symbiotic micro-organisms: (a) cow (ruminant), anterior portion only; (b) rabbit; (c) hoatzin, a bird that eats foliage; (d) plant bug ((c) after King, 1996)

modify it. The stout bodies and strong limbs of the early reptilian herbivores would have been well adapted to carrying the necessary bulk.

Tall vegetation, such as trees, presents a further obstacle. How to get at the leaves and fruits? Early vertebrates were ground-dwelling; a pelycosaur or a dicynodont could not climb a tree. They must have fed on low growing plants or young or fallen trees. None of the fodder held aloft was available to them. Later in evolutionary history, one observes the development of long necks or the ability to stand on the hind legs (bipedalism)—adaptations that would bring this food within reach.

For the small insect the plant is generally something to be lived on as well as fed on. The insect therefore has had to overcome three additional evolutionary hurdles. Firstly, they have to find the food plant, secondly, hold on to it when feeding or resting and thirdly, avoid drying out. Flight provides insects with the opportunity to scan a large area and it has been found that they can often recognize their host plant when some distance away—both smell and vision may play a part.

For an insect that lives on a plant, falling off it can often be a tragedy making it much more vulnerable to predators. Insects commonly evolve camouflage appropriate to the host plant and may be more conspicuous on the ground below. The movement necessary to get back up the plant increases their visibility and may cause them to be spotted by a bird. As well as birds there are often many other predators, such as ants, on the ground. If the insect has fallen off a tree and it is flightless there is the challenge of finding the right trunk and climbing up it. It is therefore not surprising that insects have evolved many structures to aid them in holding on to the plant, no easy feat high in a tree in a storm. Caterpillars have sucker-like 'false legs' with rows of small hooks on them (FIG. 7.12(a)). Many crickets and beetles have

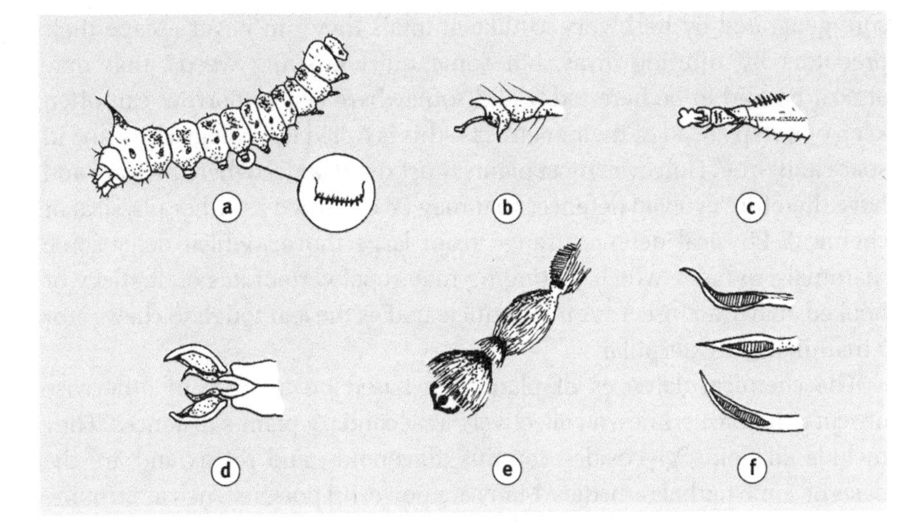

FIG. 7.12 Some adaptations of insects' feet for holding on to the plant surface: (a) moth caterpillar with small hooks (crotches) on abdominal legs; (b) lacewing fly larva with empodium (like a stalked suction cup); (c) thrips and (d) plant bug with bladder-like sacs under claws; (e) underside of leaf beetle foot showing covering with adhesive hairs; (f) adhesive hairs (after Strong *et al.* 1984)

broad pads on their feet, thickly covered in fine hairs; it has very recently been found that these excrete the smallest drop of an oily liquid that enables the tip of each hair to behave like a sucker on a smooth leaf. One beetle's resistance ability to hold-on was measured and it was found that it could withstand a pull of 80 times its own weight for two minutes. Such ability is clearly a defence against both being pulled off by a predator, like an ant, or blown off by the wind.

The relative humidity even a small distance from the surface of a leaf is quite low and, in spite of the food intake from leaves, desiccation is a further challenge. Some insects have a strongly waterproofed cuticle (skin), but there are a variety of other mechanisms. Caterpillars often drink the drops of water (guttation drops) that leaves extrude at night. Other insects also get their water directly from the plant; these are sap-suckers like aphids that are effectively plugged into the conducting vessels of the plant. Still others inject substances into the leaf to cause it to curl or form a gall so producing a humid chamber. Even crowding together in a group will reduce evaporation, though there are also other advantages to such behaviour; for example, there is safety in numbers when avoiding a predator.

Plants have of course responded through evolution to the selective pressure generated by herbivory. Unlike animals they can never escape their predators by running away, but some quick-growing weeds, that may almost be said to be here today and somewhere else tomorrow, can often keep one step ahead of their enemies—this is called by ecologists 'escape in space and time'. However, most plants must resist attack where they are and have therefore evolved defences that may be classified as either physical or chemical. Physical defences range from large thorns, which deter some mammals, to hairs, which if stinging may repel vertebrates or, if sticky or hooked, may trap insects. A thick cuticle makes the leaf tough to chew—for a mammal or a caterpillar.

The chemical defences of plants are based on a range of otherwise obscure chemicals, known collectively as secondary plant substances. They include alkaloids, glycosides, tannins, flavenoids, and resins and are the basis of some herbal remedies. Many are powerful poisons, such as atropine in deadly nightshade, and thus discourage general herbivores. However, in the course of evolutionary history, insects specializing on a particular plant have developed a biochemical mechanism for detoxifying them. Many have even gone so far as to retain these chemicals in their own bodies so as to make them distasteful or poisonous to insectivorous birds and mammals.

There are a variety of other specialized strategies that plants have evolved to protect themselves from herbivores. A particularly ingenious one involves symbiosis between a thorn bush and an ant. The bush has evolved special hollow thorns in which the ants live. They are, however, very ferocious ants and so if a mammal or another insect starts to browse on the bush the ants stream out and attack it thus discouraging further consumption of the plant!

What was the route to herbivory in insects? Undoubtedly the earliest insects fed, as do their living relatives, on decaying plant material. This applies to groups such as springtails (Collembola) and cockroaches (Dictyoptera). The ground in the Carboniferous and Permian forests must have been thickly covered with the spores and pollen from the trees. This would have provided nutritious food which is likely to have been taken by these omnivorous insects. The next step would have been to feed on the spores and pollen before these were shed, when they were concentrated in the fruiting bodies on the plants. Piercing and sucking mouthparts would have been necessary to penetrate the wall of the fruiting body and there are several, now extinct, groups of insects found in the Carboniferous and Permian that are believed to have fed in this way (FIG. 7.13). Careful studies of these insects and plant remains from these periods have been made by Russian scientists, particularly B. Rohdendorf. They have concluded that insects did not attack the living leaves of plants until at least the Permian; foliage-feeding insects are not recorded from the fossil record until the end of this period. These leaf-eating insects may have evolved this habit by a slightly different route. After feeding on plant debris and fungi on the ground, they may have started to feed on diseased parts on the plant itself, as some grasshoppers seem to prefer to do today. In both these scenarios there would have been many micro-organisms mixed with the food at an early stage and some of these associations may have led to symbiosis.

Insects being small and most plants relatively large, an insect can often obtain all the food it requires during much of its life from a single plant. This opened the way for insects to evolve a close and specialized relationship with their host plants. There is no plant so well defended that there is not some insect that has evolved to overcome the defence. Most herbivorous insects are specialists feeding on a small range of plants, usually those with similar chemistry. This has been a major factor in the evolution of so many different species of insect.

Vertebrates cannot in general be so fussy: they need a lot of food and

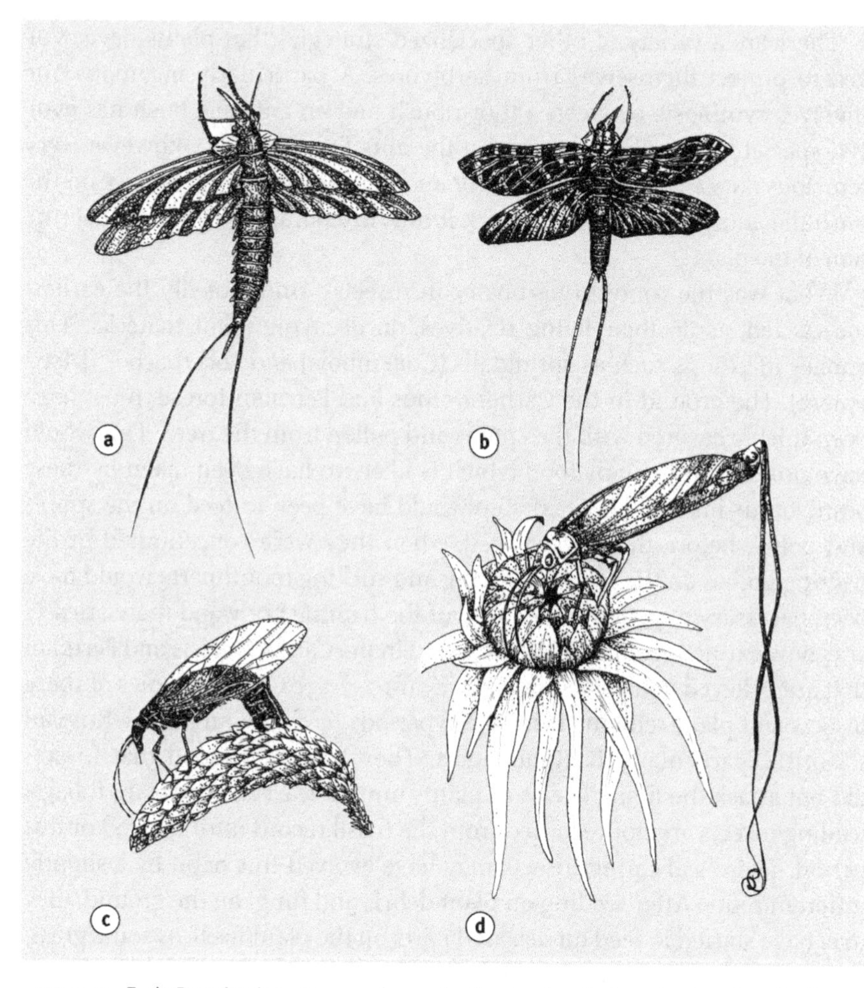

FIG. 7.13 Early Permian insects, members of extinct orders believed to be plant feeders: (a) and (b) Dictoneurida; (c) Palaeomantiscidae; (d) Diaphanopterida (after Ponomarenko in Rohdendorf and Raznitsin, 1980)

have to gather from many plants to get a meal; they cannot go searching for a single species of plant. There are some plants that have effective defences against vertebrates, as we can see from the clumps of uneaten nettles in grazed meadows. But mammals, aided by their symbionts, can detoxify a range of general plant defence chemicals. Sometimes animals are helped to locate it by the favoured food growing in large clumps—grass is an outstanding example—but most grass feeders will take other plants. Two mammals that are the supreme food specialists are the koala bear on eucalyptus (where the plant is large in comparison with the animal), and the panda whose food, bamboo, grows in exceptionally large clumps. But such specialization has its dangers if the habitat changes, and both these species are now at risk of extinction as a result of man's activities. This is a general point that has been made before: the perfectly adapted specialist is always in the greatest danger in times of rapid (in evolutionary terms) change.

There is another strategy for herbivores, which is to feed on the most nutritious parts of plants: the flowers, fruits, and growing tips. This demands mobility, achieved in insects and birds, many of which follow such a diet. For non-flying animals that are large in relation to these food sources, regularly finding enough is a particular challenge that has to be met by sophisticated behaviour. Primates have evolved such behaviour, and this need was probably one factor driving the evolution of intelligence.

Sailing into the air

Flight must have evolved in insects by the start of the Carboniferous (FIG. 7.7) and a variety of winged forms were present in the Permian (FIG. 7.13); the two pairs of wings being derived from outgrowths of the second and third segments of the thorax—the region of the body behind the head. Flight brings many advantages to insects; principal among these are the ability to escape from predators like fish, amphibians, reptiles and other insects and the potential to find new habitats for feeding and egg-laying. It is a very major asset in the exploitation of trees, permitting movement in three dimensions, flitting from branch to branch rather than being confined to walking along the twigs and branches. The downside is the danger of being blown away; insects do not have very high flight speeds, but they have evolved behaviour so that they can in general be better compared to a canoeist in a fast stream than to a piece of flotsam. In certain habitats that remain unchanged for long periods (e.g. lakes), where there is no need to move, and on wind-blown islands, where the danger of being blown away looms large, some insects have evolved so that they are unable to fly: they have lost fully-grown wings or, at least, their wing muscles. The successful evolution of flight requires the ability to take-off, to remain airborne and to land precisely: the latter seems to be the most difficult challenge.

Many theories have been advanced as to the course of the evolution of wings in insects. Until recently the three main scenarios were: gliding down from trees; allowing drift in the wind; and increasing the length of a jump while achieving a more controlled landing. However, studies of the behaviour of stoneflies have suggested a plausible alternative. The larvae of these insects live in freshwater streams and when they change into adults they swim to the surface, crawl on to stick or stone, and the adult then emerges from the larval skin. The adult needs to move away from this exposed position to the shelter of the bank as quickly as possible. It will also be an advantage if it can move up the stream before reproducing, as the larvae are likely in the course of their lives to have been swept at least a little way down stream.

It has been observed in some species of stonefly that adults launch themselves on to the water and raise their wings, which catch the wind, so they sail buoyed up by the surface tension of the water. This behaviour, termed 'skimming', is also found in mayflies which have similar freshwater-dwelling larvae (FIG.7.7(c)). It is therefore suggested that the outgrowths (gills) on the larval body, used for respiration and locomotion, were retained in some ancestral adult, where they enabled it to sail out of danger and towards a better habitat for reproduction. Every small improvement and enlargement would make its 'owner' more likely to survive and breed successfully—in evolutionary terms to increase its fitness—and so gills became transformed to wings, sailing to flying. If this is what occurred it would be an example of a common phenomenon, the evolution of something for one purpose providing a start for a new use (e.g. fins to limbs, p. 73). The novel function then opened up vast opportunities to move into new niches—new habitats, new ways of making a living.

The evolutionary advantages of the ability to move through the air are evidenced by the fact that the only group of invertebrates, other than insects, that have been really successful in the terrestrial environment are the spiders and mites and they have evolved another, albeit less controlled, method—ballooning on gossamer, a fine strand of silk.

Sharks, shrimps and sea lilies

Although several groups of marine animals became extinct or rare at the end of the Devonian, others flourished in the Carboniferous and Permian seas. Sharks were one such a group; at this time they were structurally at their most diverse. Two of the more abundant groups appear to have been predators of shellfish (molluscs) and lampshells (brachiopods) rather than of other fish. The teeth of these hybodont sharks had low crowns, while those of the holocephalians, related to the modern rabbit fish (*Chimaera*), were even more massive (FIG. 7.14(b)). Other more generalized sharks occurred in the seas, evolving to increase speed and manoeuvrability by developing paired fins which were narrowly rather than broadly attached to the body— as they were in the Devonian *Cladoselache* (see FIG. 6.3). At this time sharks were also living in freshwater, the habitat of some of the xenacanth sharks, whose teeth were characteristic (FIG. 7.14(a)); they could reach three and a half metres in length..

Most sharks and their relatives differ from the majority of other fish and amphibians in producing large eggs. Fertilization is internal and males bear claspers on the pelvic fins. These are lacking in Devonian sharks (*Cladoselache*), but their presence in groups occurring in the Carboniferous shows that the adoption of this mode of reproduction was contemporary with its evolution in tetrapods. While some produce living young, others lay eggs; each, with an attached yolk sac, is enclosed in a characteristic case secreted by the female's shell gland (FIG. 7.14(d),(e)). Called mermaid's purses, these egg cases are commonly found washed up on beaches. The young fish develops within the case and emerges as a fully formed young shark, dogfish, or skate. The cartilaginous fish have several other distinct

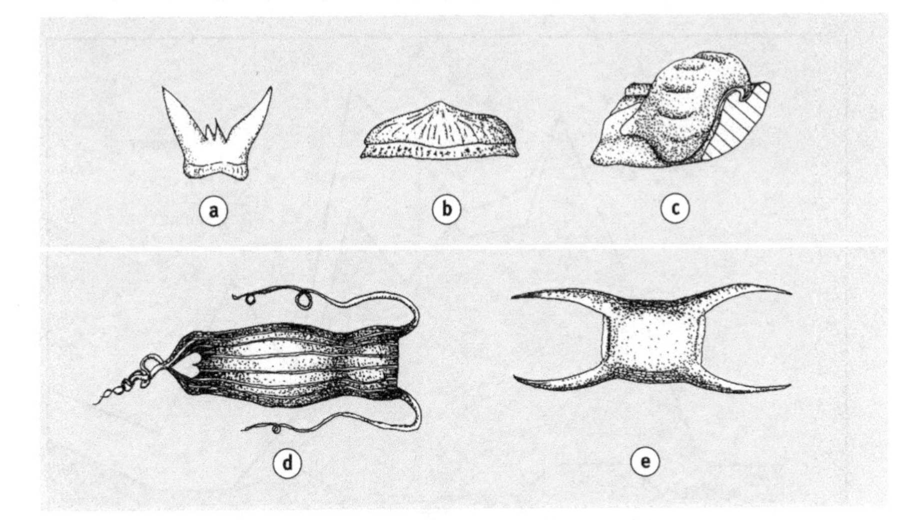

FIG. 7.14 Sharks: (a)–(c) teeth of Carboniferous/Permian forms: (a) xenacanth, (b) hybodontid; (c) holocephalid ((b) and (c) were used to crush shell fish); (d)–(e) egg-cases (Mermaid's purses) of living species: (d) dogfish; (e) skate ((a)–(c) after Janvier, 1996; (d), (e) after Goodrich, 1909)

features. For example, while most fish maintain a buoyancy neutral with respect to sea water, by means of a gas-filled swim bladder that runs the length of the body, sharks and other cartilaginous fish do this by having large quantities of oils in their livers. They are, however, usually slightly heavier than seawater and, it is generally believed, obtain lift when swimming by having the top lobe of the tail larger than the bottom lobe. When not swimming they rest on the bottom of the sea; rays do this for much of the time. Whereas most animals balance the salt in their blood to approximately that of sea water, sharks maintain their equilibrium by having a high concentration of urea, a feature they share with lobe-finned fish.

Shrimps first appear in the fossil record in the Devonian, but seem to have undergone a massive increase in diversity (radiation) in the Carboniferous, reaching their peak at the end of this period when 16 different families were present. It has to be noted that many lived in the brackish waters around the coastal plains and river deltas (in what is now North America and Europe) and conditions in such environments are more favourable to fossilization of this group than those in the sea. Thus there may have been many marine species of which we are unaware. Nevertheless, the great number of different types in the Carboniferous is striking (FIG. 7.15). The process of fossilization may also bias us when it comes to the interpretation of feeding

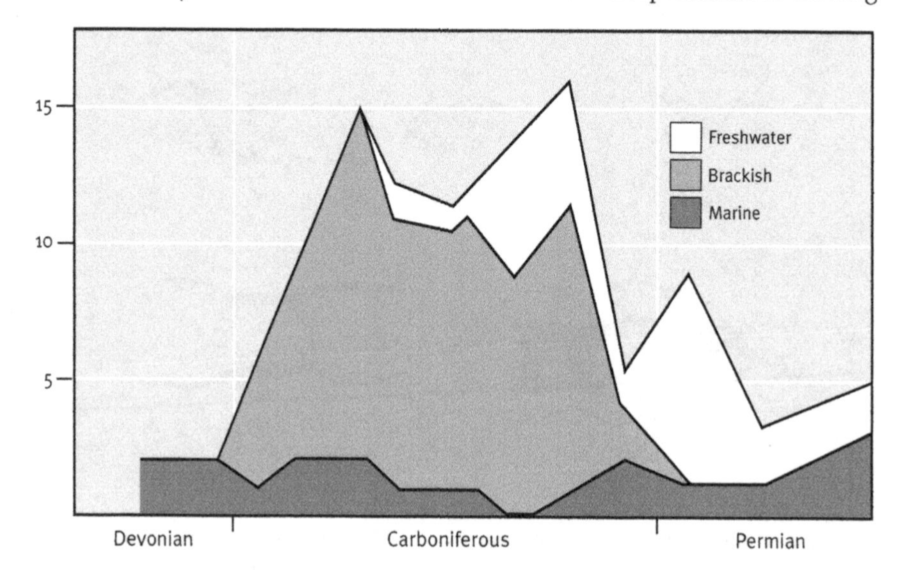

FIG. 7.15 The number of genera of 'shrimps' found in the fossil record divided according to the environments in which they are found (modified from Briggs and Clarkson, 1990)

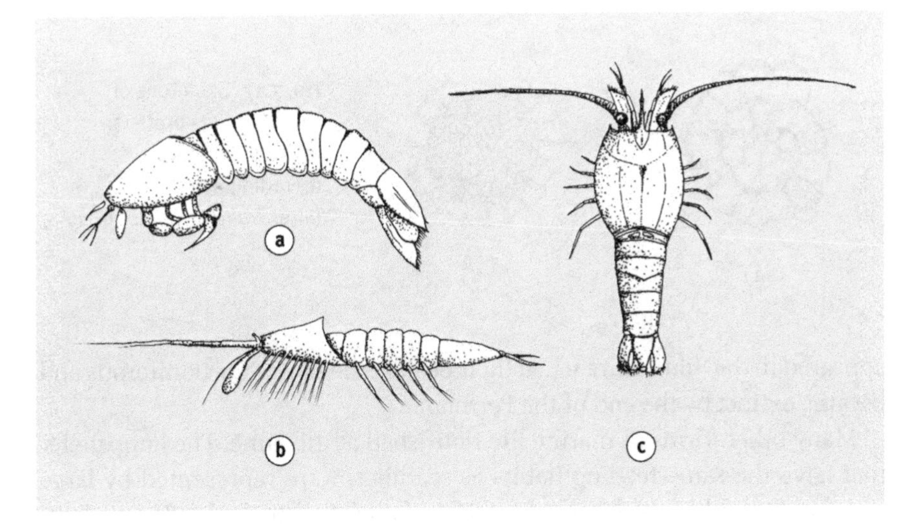

FIG. 7.16 Carboniferous 'shrimps' with different life styles: (a) carnivore (*Tyrannophontes*): (b) filter feeder (*Waterstonella*); (c) scavenger and sweep feeder (*Tealliocaris*) (after Briggs and Clarkson,1990)

habits, as the massive claws of carnivores are more likely to be retained than the fine appendages of a filter feeder. Bearing that caveat in mind, it is interesting that most of these shrimps were either carnivorous or scavengers (FIG. 7.16); the latter habit also includes low-level carnivory—consuming much smaller animals and those that are dying.

The sea lilies or crinoids first appeared in the middle Cambrian and have remained an important part of the marine fauna even up to the present, but it was in the Carboniferous that they were most successful. They belong to the same phylum as the starfish and sea urchin—the Echinodermata. Unlike most members of the group, they have a stalk (see FIG. 5.11). In some species the stalk is only present in the larval stage, but in the majority it remains throughout life and is used to attach the animal to some object on or near the bottom of the sea. Their arms, covered with fine tentacles, sweep the sea for food—algae and other small members of the plankton. These pass down the tentacles to the mouth that is at the centre of the 'flower', the animal's body, which sits in a cup. Even today, sea lilies can occur in great abundance in the deep sea, and in the Carboniferous they formed virtual underwater fields, now represented by beds of limestone largely composed of their fossils. Less abundant were the other group of stalked echinoderms, the Blastozoa, characterized by having internal respiratory tubes. They first

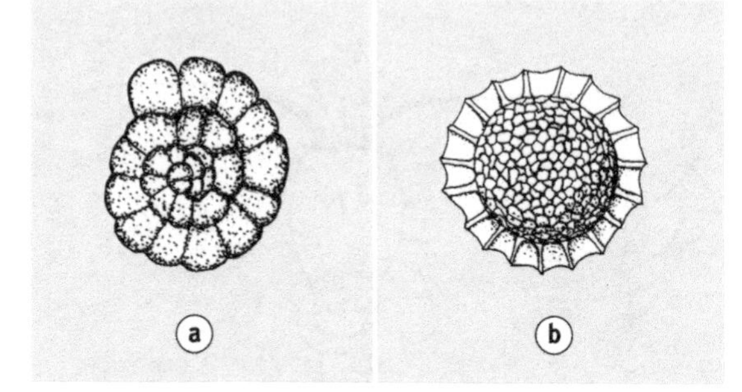

FIG. 7.17 Skeletons of carboniferous protists: (a) foraminiferan; (b) radiolarian (after *Traité de Zoologie*, 1953)

appeared in the Silurian, reached their peak in the Lower Carboniferous and became extinct by the end of the Permian.

Many other forms of marine life flourished at this time. The lampshells, that have the same feeding habits as sea lilies, were represented by large populations and many different species—some of gigantic size. These abundant filter-feeders must have had a rich brew of small organisms on which to feed. Two types of protist were particularly populous, the radiolarians and foraminifers (FIG. 7.17). The former have a skeleton of silica, the latter a shell of calcium carbonate, both features that remain after the organism has died, and form sediment that may eventually become rock. Foraminifera were so abundant in the Cretaceous that their shells contributed to the massive deposits of chalk laid down at that time.

This is the one period in history, since the Cambrian, when the types of organism that build framework reefs were scarce. The stromatoporoid sponges and the tabulate corals became rare at the end of the Devonian; scleractinian corals—so important today—had not yet evolved. Reef mounds were however constructed by calcareous algae and sponges, with other organisms, rugose corals, sea mats (Bryozoa), sea lilies and lamp shells, that attach themselves to the mound, adding to the accumulation and interception of material. These reefs could become very large as can be seen from the one now exposed as El Capitan near Signal Park, Texas.

The great Permian extinction

This was the largest extinction in fossil history; estimates of the proportion of species of marine animals that became extinct are around 95 per cent and over half the families were lost. Among the groups that became extinct were the trilobites, the rugose corals, and the blastozoan echinoderms. Only a few radiolarians, ammonoids, lampshells and sea lilies survived and the two

two last-named groups never achieved the same diversity again. On land the dominance of the labyrinthodont amphibians and the early reptiles was ended; the pelycosaurs seem to have gone before the end of the period. The widespread *Glossopteris* forests appear to have been destroyed and some groups of seed ferns disappeared. For the only time in fossil history a substantial number (eight) of orders of insect became extinct.

A few groups seem to have sailed on relatively unscathed, in particular the sharks and other fish, together with debris-feeding foraminifers and various molluscs (both snails and bivalves). Many of these invertebrates were bottom-dwelling (benthic) species that could probably tolerate a lack of oxygen, but how and why the fish largely survived is a mystery.

This extinction is actually two events separated by about eight million years. Some groups, such as the ammonites, that were greatly reduced by the first event radiated, with many new species evolving, before the second event. This occurred around 248 Mya and marks the end of the period of geological history extending from the Cambrian and known as the Palaeozoic. The second event had more extensive effects and although there is still much uncertainty, it is better understood than the first. By no means all experts would agree with the scenario put forward here (FIG. 7.18), but I find the present evidence persuasive. It is not surprising that such a unique event was due to the chance synchronization of a variety of causes.

The key event of this period was the eruption of the Siberian Traps; it has recently been suggested that an asteroid also hit the earth at this time; this may have been the trigger for massive sheets of basalt spewed up from the mantle of the earth. It is calculated that there were

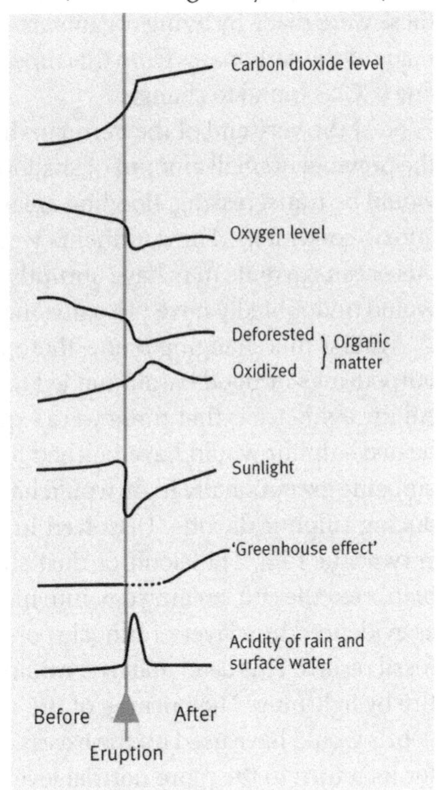

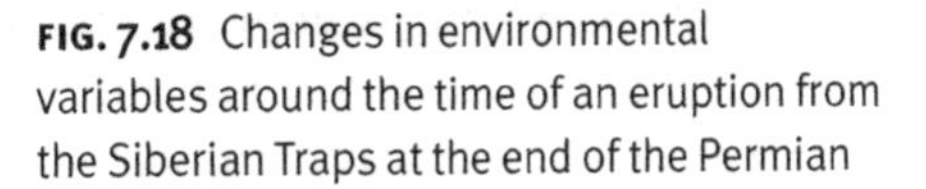
FIG. 7.18 Changes in environmental variables around the time of an eruption from the Siberian Traps at the end of the Permian

at least two million cubic kilometres of liquid rock extruded over a period of less, possibly much less, than a million years. This happened in a world where carbon dioxide levels were low and oxygen levels high (see FIG. 7.1). The low level of carbon dioxide would continue the reverse greenhouse effect that had been operating from perhaps the middle of the Carboniferous. This would lead to further accumulation of ice in the polar regions causing a fall in the sea level. The 'squeezing' together of the continents to form Pangaea would also contribute to the fall in sea level; it has been calculated that at this time only 13 per cent of the continental shelves remained covered with water. This fall in sea level exposed organic sediments that would have oxidized particularly rapidly given the high level of atmospheric oxygen. This would have released proportionally more of the lighter carbon isotope (^{12}C) that had been previously locked up in the sediments as these were made by living organisms. This is why today, when an analysis is made of the carbonates from this time, the ratio of this isotope to the heavier one (^{13}C) is found to change.

So at the very end of the Permian the climate was becoming warmer and the previously small amount of shallow sea was actually increasing. The sea would be transgressing, flooding areas previously dry, which often leads to anoxic conditions. The continents were finally suturing together and various ocean currents may have abruptly changed direction. All these changes would undoubtedly have put some marine species under pressure.

Against this changing scene, there were a series of eruptions with great outpourings of flood basalt that led to the formation of the Siberian Traps. All the evidence is that these were explosive events. Large clouds of superheated sulphur would have belched forth and, with the oxygen level in the air being exceptionally high, would have immediately burst into flames producing sulphur dioxide. Dissolved in water this would have descended as very acidic rain. The cloud of dust shot up from the volcano would have blanketed the sun, turning day into night. Large amounts of vegetation died as evidenced by a layer of fungal spores—a most unusual occurrence in the fossil record. This dead material would have been vulnerable to being set on fire by lightning. The burning of the sulphur and of the carbon in the plant debris would have used up the excess oxygen in the atmosphere accounting for its return to the more normal level (see FIG. 7.1). Although much of the dust would have settled, allowing the sun to shine on the earth again, there would have been a period when fine dust reduced the strength of the radiation and the climate was extremely cold, causing more ice to form and the seas to regress again. However, the dust would have eventually cleared and the earth would have experienced the full strength of the sun's radiation again; but unlike the situation before the eruption, there would now be a shield of carbon dioxide and other greenhouse gases so radiation would be trapped and temperatures would have risen dramatically; the ice sheets would have melted and sea levels risen sharply, bringing anoxic waters from the seabed to the surface.

This sequence of events would have happened over what in geological terms would be a very short interval of time, but is most likely to have been repeated several times. Some eruptions would have been colossal with a global effect; others may have been less violent with a more limited effect. But overall, throughout this period of probably hundreds of thousands of years, life would have been continually challenged with alternating hot and cold periods, falls and rises in sea level, bursts of strongly acidic rain and frequently anoxic seas and oceans.

There is less evidence related to the first extinction event that came some eight million years earlier. It may have been similar, though perhaps on a shorter time scale. The Emeishan Traps in China are another area of flood basalt and they are believed to date from the right time—256 Mya. It is easy to see that, with these two events and the series of eruptions within each, not all species would have succumbed at the same time. The wonder is perhaps not that so many species became extinct, but that any survived at all.

CHAPTER 8

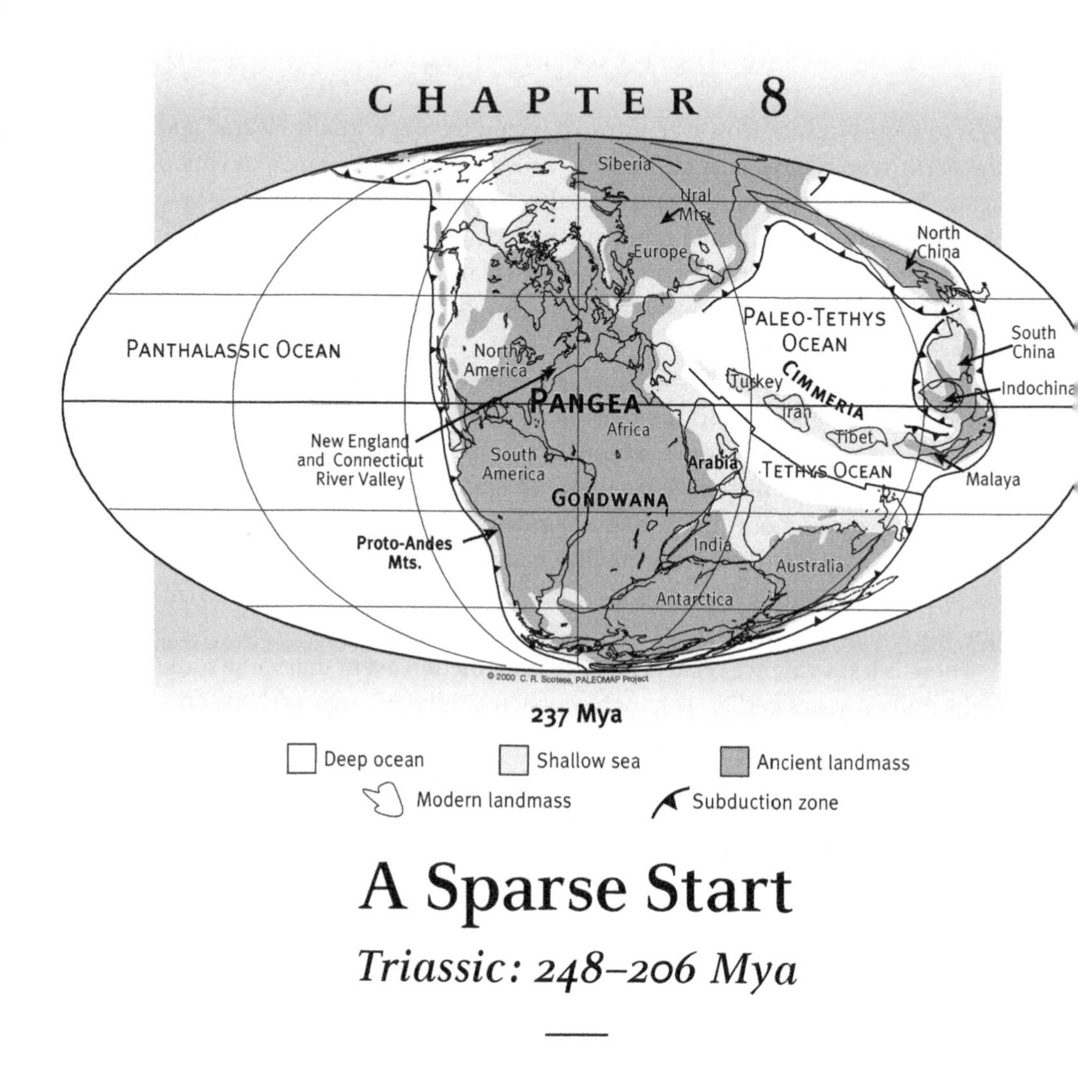

A Sparse Start

Triassic: 248–206 Mya

THE supercontinent Pangaea remained intact throughout this period and various small land masses that now constitute South-East Asia moved to join the eastern corner of Laurasia. Although Pangaea stretched from near one pole to near the other it seems that there was no land that was actually over a pole and there is no evidence of continental ice sheets. The ocean currents will have flowed freely, particularly in the enormous Panthalassic (Palaeopacific) Ocean, bringing warm water to colder latitudes. The Tethys Ocean existed as a deep bight, at tropical level, to the eastern side of Pangaea. The sea rose from its low level at the end of the Permian, so that along the margins of the Tethys Ocean there were often luxuriant tropical swamp forests. The period was probably a relatively warm one, gradually becoming hotter; away from the oceans conditions were often very arid and there is evidence of the evaporation of inland seas in the large deposits of rock salt, gypsum, and anhydrite formed at this time. Animal and plant life seems to have been relatively sparse when the period began.

Ferns and conifers

At the start of the Triassic period fern spores are widespread in the fossil record. Ferns are often the early colonists of the rather barren landscapes that are likely to have been extensive during the final phases of the Permian extinction and throughout the early part of the Triassic. Although the trees often belonged to other groups, the predominant plants at ground level were ferns. This was an 'age of ferns'.

From about the beginning of the Middle Triassic the seed fern (*Dicroidium*) spread from tropical locations and became widespread throughout Pangaea. This cosmopolitan distribution of flora (and fauna) is characteristic of much of the period, a reflection of the lack of sea barriers to prevent dispersal. Though the conifers and their relations were first found in the Carboniferous it is throughout the Triassic that they become more frequent and more diverse in the fossil record. Their sunken stomata and thick leaf coverings, adaptations to dry conditions, would have been important advantages in the many arid parts of Pangaea. With two exceptions, all the major groups found at this time, or early in the Jurassic, are extant today. The families containing the yellow woods (Podocarpaceae) and the monkey puzzle and Norfolk Island pine (Araucariaceae) (FIG. 8.1) both flourished in the southern part (Gondwana) of the supercontinent; later they became more widely distributed, but today their natural distribution is restricted to their old headquarters—the southern hemisphere. In various parts of Pangaea fossils have been found that show the basic features of pines (Pinaceae), redwoods (Taxodiaceae), cypresses (Cupressaceae), and yews (Taxaceae), though all these trees become more widespread and abundant in the periods covered in the next chapter. With the exception of the giant redwoods, they remain widely distributed today although normally ceding the dominant position in tropical and warm temperate forests to the more recently evolved angiosperms (flowering trees) (see p. 163).

The conifers described above all belong to a larger group called gymnosperms, which literally means 'naked seeds', for the seeds, at the time of pollination, are not enclosed in any tissues of the parent; though this may happen subsequent to fertilization. There were two groups of gym-

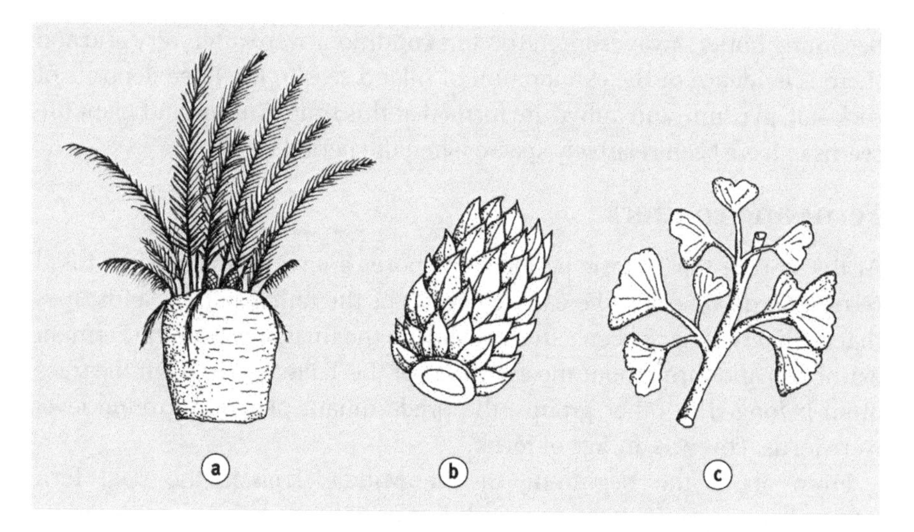

FIG. 8.1 Modern representatives of Triassic trees: (a) Cycad with cones; (b) shoot of a monkey puzzle (*Araucaria imbricata*); (c) shoot of maidenhair tree (*Ginkgo biloba*)

nosperms, in addition to the conifers, that were common in the Triassic flora. These were the cycads or royal palms (Cycadophyta) and the maidenhair trees (Ginkgophyta) and representatives of these are living today (FIG. 8.1). There are over a hundred species of cycad found in tropical and subtropical regions, mostly in dry locations and some at least live for over a thousand years. Although today's plants are large, from about three to eighteen metres, both smaller and larger cycads probably occurred in this period, although there is considerable uncertainty about the exact structural interpretation of the various pieces that have become fossilized. The plants are of different sexes; the males produce abundant pollen. Certain weevils have been found to live in the male cones, where they get covered with pollen; as they also visit the female cones and so transfer the pollen, this may represent the earliest evidence of the pollination of plants by insects. Modern cycads have very toxic juice protecting them from most herbivores but, in addition to the weevil, there is one rather unusual family of plant bugs that feeds on them, and on them alone. The different families of plant bugs mostly evolved 100 to 200 million years ago, so this too is probably a long association. Because plant chemistry plays an important part in such associations it also suggests that the toxicity of cycads may be an ancient feature. During the next geological period, the Jurassic, cycads were a very major component of the flora, but one can only speculate how this affected herbivorous dinosaurs.

The maidenhair tree, *Ginkgo biloba* (FIG. 8.1(c)), is the only surviving member of its group, which was at one time widely distributed. It probably owes its survival over the last several hundred years to being carefully cultivated in temple gardens in China and Japan and is now grown in many towns in temperate parts of the world; it is very resistant to pollution. One is tempted to guess that this ability may have assisted in the survival of the ancestors of this and other maiden-hair trees through periods when sulphur dioxide and other pollutants were produced by the massive volcanoes at the end of the Permian. The trees are of different sexes; the female trees shed large numbers of fruits at the same time, consisting of a round nut covered with a fleshy outer layer. To humans this is foul smelling, but probably served to attract some reptiles or mammals that would distribute the seeds. Fossilized leaves very similar to those of the extant tree are known from 80 million years ago, but those ginkgoales living in the Triassic period were probably very different in appearance from today's tree.

Becoming more like mammals

As was mentioned in Chapter 7 three great groups of reptiles can be recognized. It is from one of these, the synapsids, that mammals eventually evolved. At the beginning of the Triassic there were various representatives that had survived the Permian extinction. The herbivorous dicynodonts (FIG. 7.11) had become less diverse, but one, *Lystrosaurus*, was extremely widely distributed in the Early Triassic and was perhaps the most successful member of the group, though its abundance lasted for only a short period in geological terms. The largest specimens measured about two metres in length. A number of different species of dicynodont are found in the Middle and Upper Triassic and it is possible that some of these may have been able to stand on their hind legs to browse trees and shrubs, cutting through the twigs and foliage with their horny tortoise-like beak. The dicynodonts died out at or near the end of the Triassic.

Another group that spanned the Permian–Triassic border are the Therocephalia. They are particularly interesting in that, in the Permian, some of them were the earliest insectivores and at that time they exhibited a wide range in feeding habits and in size (30 to 200 centimetres). In the Lower Triassic they were smaller and were mostly insectivores (FIG. 8.2(a)).

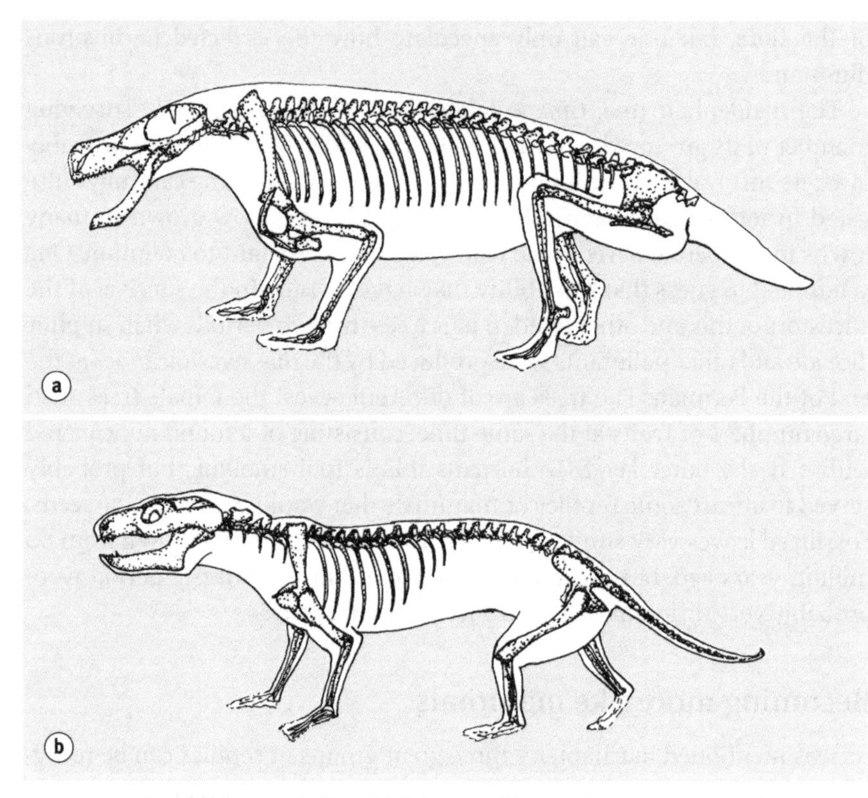

FIG. 8.2 (a) therocephalian, *Ericiolacerta*; (b), cynodont, *Massetognathus* (after Kemp, 1982)

The cynodonts probably evolved from the ancient therocephalian stock. They first appeared in the Late Permian and flourished and diversified in the Triassic gradually showing more of the features that characterize mammals. Probably initially insectivores, they also evolved into herbivores and carnivores. They varied in size, but mostly ranged, like dogs, from 50 to 150 centimetres (FIG. 8.2(b)).

The early members of this group, and earlier tetrapod groups, were sprawlers (FIG. 8.3). That is they walked, as a newt or salamander does, with the legs bent at the elbow and knee, and the feet well out from the body. Stride length is limited and this may be compensated for by swinging the shoulder from side to side with each step. When the body swings like this the lungs are squeezed, so these amphibians and reptiles cannot run and breathe at the same time. If you watch a lizard, it runs for a short distance and then stops and breathes. However, in the cynodonts there was an evolu-

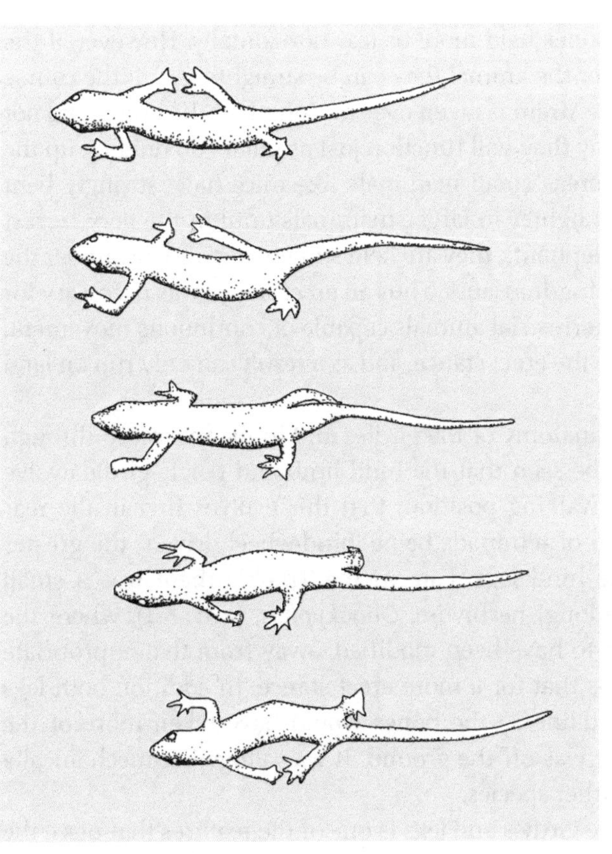

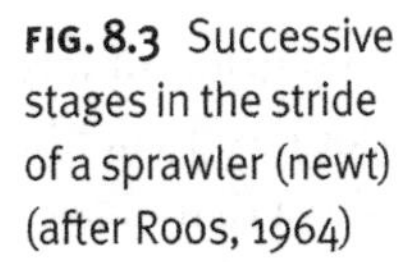

FIG. 8.3 Successive stages in the stride of a sprawler (newt) (after Roos, 1964)

tion towards the erect stance one sees in mammals. In this stance the anterior of the body is suspended from the shoulders, so breathing and movement can be carried out by different sets of muscles. Air is pumped in and out of the chest cavity by the muscular diaphragm and in many mammals this synchronizes with movements of the backbone during running so that breathing is enhanced. But the erect stance also allows other evolutionary developments.

It brings the elbows and knees and hence the feet, under the body; this increases agility and speed, two attributes very important in survival for escape from enemies and the capture of prey. Furthermore, as already pointed out (p. 76), if we imagine enlarging an animal by keeping the same proportions, its weight goes up by the cube, while the strength of a bone increases by its cross-section, the square. If limbs are bent at the elbows and knees there is a turning moment that, as it became larger with increased size, would snap the bones held more or less horizontally. However, if the limbs are directly under the animal they can be straightened, in the course of evolution, so that the strain is taken over the length of the bones and not across them: in this way they will function just as pillars do holding up the roof of a church. Whereas small mammals like mice have strongly bent limbs, they become straighter in larger mammals until in the very largest terrestrial mammals, elephants, they are held straight and directly under the body. The same is true for dinosaurs. Thus an erect stance was necessary for the evolution of large terrestrial animals capable of continuous movement. Crocodiles do not have the erect stance, and as a result can only run on land for short distances.

When the detailed anatomy of the girdles and limbs are traced through the cynodonts, it can be seen that the hind limb and pelvic girdle evolve to allow a semi-erect walking position; that this evolves first in the rear quarters is a reflection of tetrapods being 'hind-wheel driven', the greater strength being in the hind legs. Late in the Triassic there was a small (about 55 centimetre long) herbivore, *Oligokyphus*, (FIG. 8.4), where the front joint also seems to have been modified, away from that appropriate for a sprawler, towards that for a more erect stance. In addition both legs were closer to the mid-line so the bones would have taken more of the weight when the belly was off the ground. It was altogether mechanically more efficient than earlier species.

This evolution of the girdles and legs is one of the features that place the cynodonts on the evolutionary route that leads to mammals. Another mammal-like character is the way the teeth are differentiated into incisors and canines used for food-gathering, with those behind for crushing and chewing. If they chewed they probably had a muscular tongue to manipulate the food. The bone arrangement in the ear is not quite like that of mammals, but they could undoubtedly hear. The shortening of the ribs in the lumbar

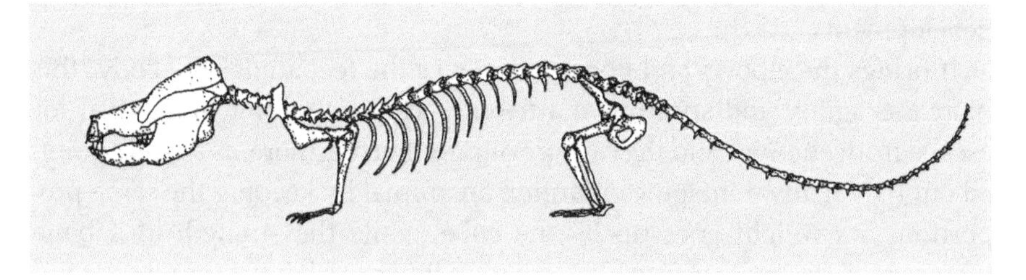

FIG. 8.4 Late Triassic cynodont, *Oligokyphus* (after Kemp, 1982)

region suggests that they probably had a diaphragm, which—as mentioned—would allow breathing to be regular and independent of other body movements. Tom Kemp of the University of Oxford who has made a detailed study of this group considers that the evidence for them being warm-blooded is very strong.

FIG. 8.5 Early mammal, *Megazostrodon* (after Charig, 1979)

Thus we can see that the cynodonts, particularly that group known as the Cynognathoidea, were the protomammals; they become scarce in the fossil record in the Upper Triassic at about the time that the earliest mammal remains are found. These are small, about mouse-sized (up to about ten centimetres) (FIG. 8.5). Today such small animals are mostly nocturnal and it is speculated that these early mammals would have been most active at night, in which case they must have been warm-blooded.

On having warm blood

Although most animals can easily be designated cold-blooded and warm-blooded, there are many intermediates. Some fish, like barracuda and sharks, have an arrangement of blood vessels that keeps their muscles warm; lizards and some insects raise their body temperatures by sunning themselves; and other large insects, like bumble bees and hawk moths, often have to vibrate their flight muscles to warm them before flying. Warm-blooded animals normally hold their body temperatures constant, but they will lower this temperature when in a state of torpor. Bats may do this during the day and, when disturbed, 'shiver' briefly before flying off. Some hummingbirds become torpid at night when the body temperature will fall with the air temperature until this reaches about 20 °C; if the external temperature falls further, the birds maintain their body temperature at 20°C. In animals that hibernate, such as bears and hedgehogs, body temperatures may fall to as low as 6°C; while in the arctic ground-squirrel it falls to, but is maintained at, a remarkable –2.9°C (the tissues are protected by the equivalent of 'antifreeze'). When these hibernators wake up special reserves of a brown fat are 'burnt' to raise the body temperature. From these examples it can be seen that the distinction is not really dependent on the blood temperature, but is between those animals whose body temperature is largely determined by the external environment and those that generate heat internally and have the ability to hold their temperature at a certain value.

The two groups of living animals that maintain a stable body temperature are birds and mammals. The normal temperature varies from group to group being in the ranges 38–41 °C in birds and 35–38 °C in most mammals. Heat is produced by the muscles when active, by shivering (involuntary muscular activity), and by the organs of the body (gut, brain, lungs, liver, etc.). In ourselves, the latter are the source of over two-thirds of the heat produced when we are resting. Warm-blooded animals conserve heat by a covering of fur or feathers and layers of fat (blubber) under the skin. They lose heat by evaporation—by sweating and/or panting—and simply by radiation and convection from the surface.

A small animal will have a larger amount of surface in relation to its volume than a large animal. (This arises simply from the mathematical relationships of the surface area to the volume of a sphere). A small lizard will cool down much more quickly than a large crocodile. Large animals will have such a considerable volume of organs and muscles in relation to their surface area that keeping cool will be a more significant challenge than keeping warm. In this respect it is notable that large tropical animals—elephants, rhinoceroses, and hippopotami—lack a covering of hair, though their now extinct relatives, such as the mammoth and woolly rhinoceros, that lived in cold climates were hairy. Whales and the majority of adult seals have also lost most of their hair, but this is associated with streamlining for an aquatic life and they are insulated by a thick layer of blubber.

Being warm-blooded, and so maintaining a stable temperature, has a number of advantages. Firstly, it enables the brain to rely on a standard response from its component neurons and hence allows the evolution of complex brains in the wildly fluctuating temperatures of terrestrial habitats. This permits a good speed of response to food or danger and the resulting movement can be maintained independently of the weather; there is no warming-up time. Secondly, it increases the growth-rate of the young, both before birth and after. What we may term the short adolescence allows the development of parental care without taking up too much of the parent's life-span and so curtailing the production of further young. If cold-blooded crocodiles were to care for their young until these reached adult size they would be able to reproduce only every nine years or so, but most birds and mammals have one or more broods per year. The mortality of their young is reduced by parental care; there is also the opportunity of learning from the parent's behaviour. There are some cold-blooded animals that show limited parental care, ranging from earwigs to certain fish, frogs, crocodiles, and snakes, but their care is short-term and does not extend until the young are well on the way to being full grown, as it does in most mammals and birds. Not only can warm-blooded animals care for their young as they grow, but they can provide conditions in a nest or similar refuge that are more suitable than those of the external environment, and in temperate latitudes this extends the breeding season. Again, there are exceptional cold-blooded animals that achieve this by special adaptations: for example the diamond python maintains temperatures in its nest several degrees above the surroundings. However, the warm-blooded animals have achieved overall better survival and a shorter generation time. These together constitute a higher intrinsic rate of population increase, a key component of evolutionary success.

In the Triassic there were numerous large predators that would have preyed on the smaller reptiles and early mammals. One way of reducing the risk of being killed and eaten would have been to be nocturnal and, although there are now specialized nocturnal predators (e.g., owls), most very small mammals are still nocturnal. However, because of their large volume-to-surface area ratio, the large predators would have retained their body heat, built up during the warm day, longer than the small prey. The latter would quickly become less alert and slower when the temperature fell, as it does on land, at night or in cloudy conditions; they then literally become sitting targets for the still warm larger predators. The need for small animals to remain fully alert and active whatever the environmental conditions would have provided the main evolutionary driving force for the development of warm-blooded physiology on land. The more constant temperature conditions in the sea (p. 78) did not provide the stimulus for this evolutionary step.

The dawn of the dinosaurs

We have followed the evolution of one of the great groups of reptiles, the synapsids, to the mammals. Another group, the diapsids, contains the evolutionary roots of the dinosaurs, birds and all living reptiles except the turtles and tortoises. During the Triassic period there were a number of types that flourished for a time, but died out at or before the end of the period; there is also evidence from the fossil record that the main groups of dinosaurs first appeared in the Late Triassic.

The powerfully 'beaked' rhynchosaurians (FIG. 8.6(a)) were the dominant herbivores for much of the middle portion of the Triassic and were particularly associated with the seed fern, *Dicroidium*. Michael Benton, of the University of Bristol, who has studied them in detail has suggested that the disappearance of this plant, which was their food, caused their extinction in the Late Triassic. However, large herbivores are not usually so specific and it may be that the change in the environment, whatever that was, that lead to the demise of the seed fern had a similar effect on the rhynchosaurs. This group is considered to be an early off-shoot of the group known as the archosaurs to which the other reptiles discussed below belonged.

The aetosaurs (FIG. 8.6(b)) were a group of heavily armoured herbivores that were found only in the Late Triassic, as were the related carnivorous phytosaurs (FIG. 8.6(c)), that were crocodile-like in appearance and probably in habits, reaching lengths of nearly four metres. Other carnivorous reptiles, including the ancestors of modern crocodiles, were present in the late Triassic and some had relatively long legs, suggesting that they were capable of moving at speed.

The fossil record of the Late Triassic also reveals several types of reptile that can be regarded as true dinosaurs. The predominantly herbivorous prosauropods (FIG. 8.7(a)), with their relatively long necks, were often bipedal, moving mainly on their hind limbs. This stance and their long necks would have enabled them to browse on trees and tall shrubs, food that could not have been reached by other herbivores such as the aetosaurs. In spite of their name they were not the ancestors of the great dinosaur group, the Sauropoda, but appear to have been an evolutionary dead end.

The groups (Saurischia and Ornithischia) to which the many dinosaurs of the Jurassic and Cretaceous belong appear in the Late Triassic, but do not seem to have been dominant or abundant. For them this was a 'fuse' period (p. 44). They were an insignificant group prior to their radiation after the extinction of the existing dominant groups. *Coelophysis* (FIG. 8.7(c)) is a delicate and relatively small predator, about one metre tall and two and a half metres long; it had sharp pointed teeth and probably fed on small reptiles and insects; however it belongs to the theropod group of the Saurischia of which the giant Cretaceous *Tyrannosaurus* is widely known. Representatives of the bipedal herbivorous ornithischian dinosaurs are also

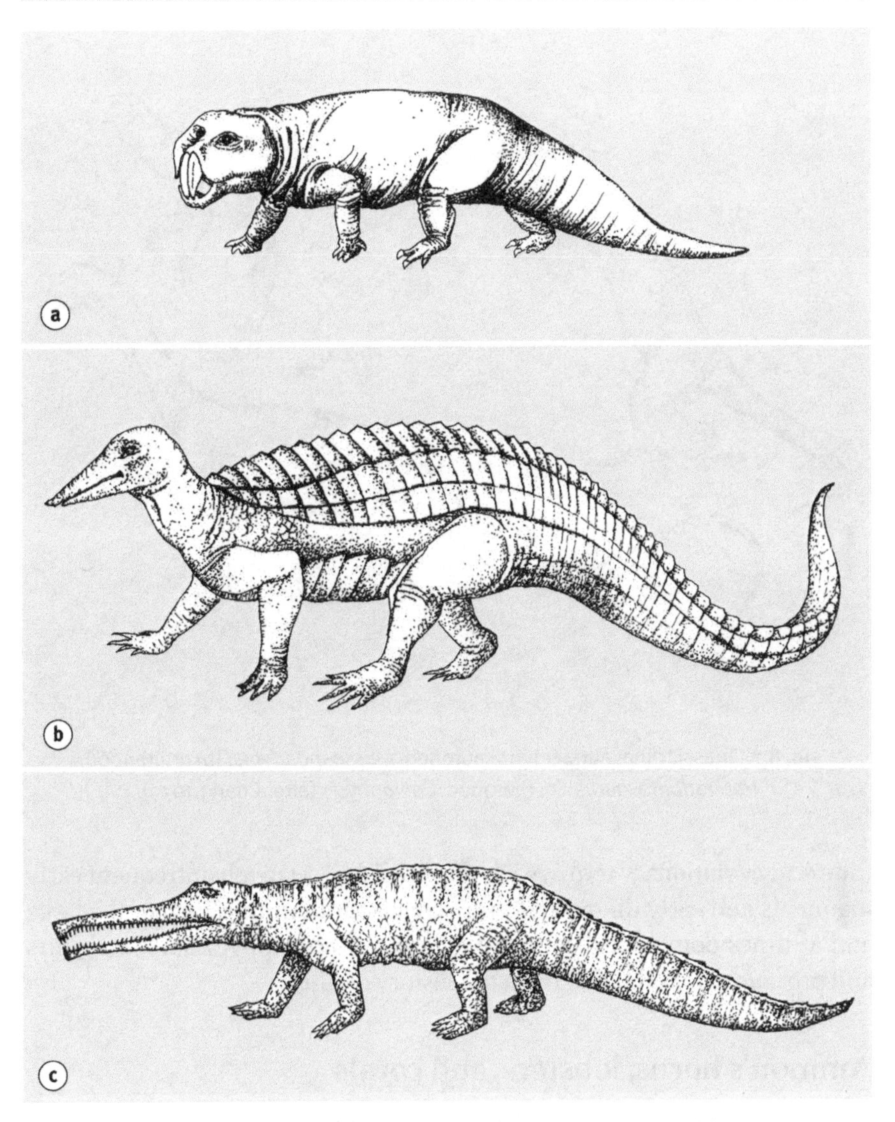

FIG. 8.6 Triassic reptiles that are not dinosaurs: (a) rhynchosaur; (b) aetosaur; (c) phytosaur. ((a) 1.3 m, (b), (c) 3m)
((a) and (b) based on Benton, 1983; (c) after Charig, 1979)

found in the Late Triassic; they were small animals, less than a metre in length, including the tail (FIG. 8.7(b)), and probably relied on speed and alertness to escape their predators.

So we can see that those species in the Triassic with the most glittering

FIG. 8.7 Triassic dinosaurs: (a) prosauropod, *Massospondylus*; (b) ornithopod, *Heterodontosaurus*; (c) theropod, *Coelophysis* (after Charig,1979)

future, in evolutionary terms, were the small and relatively infrequent early mammals and early dinosaurs (saurischians and ornithischians); the large and at times dominant dicynodonts, rhynchosaurs, phytosaurs, aetosaurs and prosauropods all faded from the history of life.

Ammon's horns, lobsters, and corals

The spectacular coiled shells of the ammonoids have been a conspicuous feature of marine life since the Devonian (FIG. 8.8); they gain their name from a supposed resemblance to the horns of the Egyptian god Ammon. They belong to that group of the Mollusca, also containing the squid and octopus, known as the Cephalopoda. All living members propel themselves with a jet of water from a musculature tube—the funnel, though there is some doubt as to whether this was the method of all, or indeed any, of the ammonoids. Cephalopods have well-developed eyes, brains and nervous

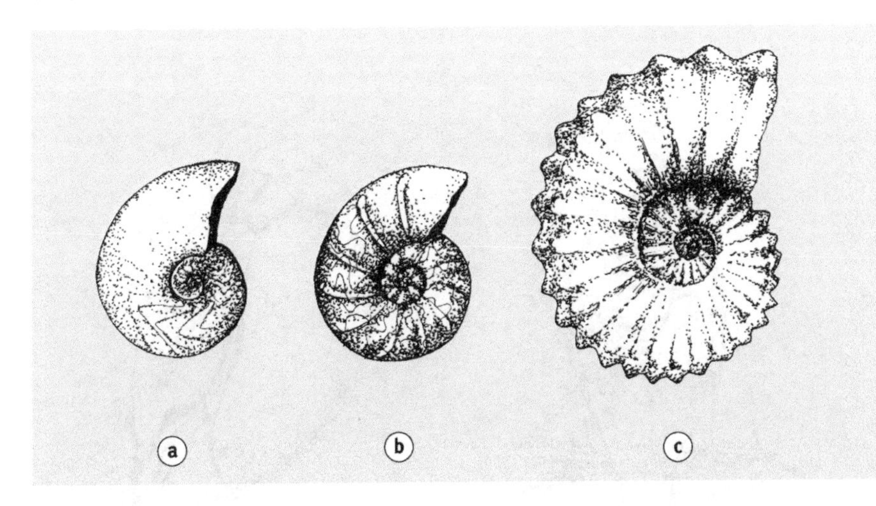

FIG. 8.8 Ammonoids showing differences in sculpturation: (a) *Manticoceras* (Devonian); (b) *Ceratites* (Triassic); (c) *Acanthoceras* (Cretaceous) (after Moore, 1957)

systems rivalling in efficiency and complexity those of many vertebrates. As predators and scavengers they occupy the same niche as the fish, which today have become dominant; but it has not always been so. Cephalopods have been very successful in various periods and one of the striking features of the ammonoids is the diversity of species (thousands), many of which are present in the fossil record for only a limited time; therefore they are widely used to identify the strata in which they are found. A quarter of the known genera evolved in the Triassic, but they were reduced to near extinction at the end of the period only to radiate profligately again in the Jurassic and Cretaceous, prior to their final extinction.

Among animals there are, in addition to sheer size, four antipredator defences: physical (armour—a strong shell), chemical (poisonous—distasteful), visual (camouflage or hiding in a shelter) and rapid movement. Strong shells and speedy movement are not easily compatible. Among the cephalopods, the ammonoids and the nautiloids (FIG. 8.9), which also have a long fossil record, shared the defensive strategy of a strong shell; the squids in contrast have gone for speed and camouflage—squirting a 'smokescreen' of ink.

Predators often also need speed to capture their prey so it is not surprising that the shells of many of the early ammonoids had fine sculpturing on the shell that would reduce drag (the dimples on a golf ball also do this).

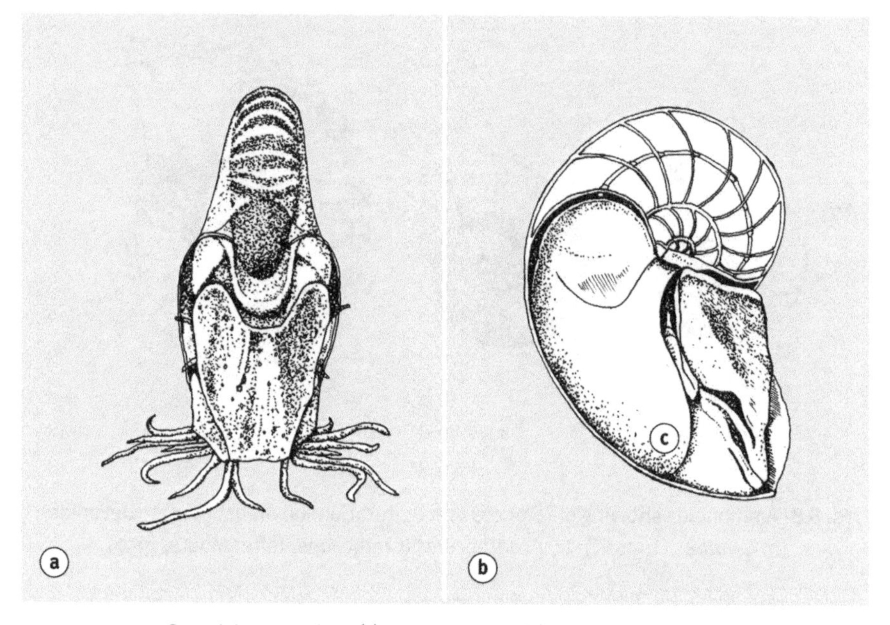

FIG. 8.9 Living nautilus: (a) anterior view; (b) longitudinal section to show chambered shell and other features of the internal structure ((a) after Willey, 1902; (b) after Sedgwick, 1898)

However, from the Triassic onwards the width of the ribs, relative to the diameter of the shell, increased: the shells were rough and strong and would have been detrimental to swimming (FIG. 8.8). The prime evolutionary pressure seems to have been for protection against shell crushing predators. The ammonoid shell was chambered as it is in the living nautilus. In the latter, these chambers are filled either entirely or partially with liquid or gas, the latter serving to neutralize the weight of the shell. The nautilus varies the proportions of gas and liquid to control buoyancy and its position in the ocean. Nothing is known of the soft parts of the ammonoids, but horny, and sometimes mineralized, jaws and a radula (a ribbon-like tongue) have been found in a few fossils.

In many species there were larger and smaller forms, and these are believed to have been females and males. Such differences in size in living cephalopods is associated with complex courtship rituals. Ammonoids laid numerous small eggs that hatched into 'ammonitella' only a millimetre or two in diameter. Their small size suggests that they were planktonic, living in the surface waters of the seas. The adults of most species frequented coastal waters and had shell diameters from 10 to 60 centimetres, but some reached gigantic sizes (with a shell diameter of over two metres). These habitats—surface and shallow waters—are those greatly affected by many of the factors considered responsible for the mass extinctions. Ammonites, particularly prone to extinction, finally disappeared from the fossil record at the end of the Cretaceous, like many other components of the plankton.

The nautilus, in contrast, lives in deep water where it is subject to great water pressures. It is therefore not surprising that its shell has been found to be resistant to implosion withstanding pressures of 80.5 kilograms per square centimetre, which is equivalent to being 785 metres deep in the sea. It can tolerate low levels of oxygen and feeds on carrion; unlike most other living cephalopods it grows slowly, and lives for as much as 12 years. In further contrast to ammonites, the nautilus, like other living cephalopods, lays only a few large eggs. The many nautiloids in the fossil record are likely to have had different features from today's species which are highly specialized for a particular way of life, but one that probably made them less vulnerable to extinction events.

There are two groups that are conspicuous animals in today's seas which appeared for the first time in the Triassic. First, the lobsters and their allies (Decapoda), which have five pairs of legs on the thorax; and second, the scleractinian corals, the supreme builders of framework reefs. Neither group radiated significantly until the Jurassic, another example of groups being relatively inconspicuous in one period and coming to great importance in the next.

The scleractinian corals are the dominant reef-builders from the Jurassic onwards, although calcareous algae and sea mats (Bryozoa) also contribute. They are restricted to warm subtropical or tropical seas and most species have a symbiotic relationship with one of the algae of the group zooxanthellae, which are mainly responsible for their bright colours. The algae photosynthesize and therefore they can live only in clear water, whilst the coral animal, the polyp, catches small zooplankton with its tentacles (FIG. 8.10). There is some exchange of nutrients between the carnivorous polyp and the photosynthesizing alga; the relative importance of the components varies from coral to coral. Reference has already been made (p. 53) to the importance of reefs in providing a great amount of varied ecospace and coral reefs are the most diverse of marine ecosystems. Although it is now recognized that they have evolved an ability to recover from natural disturbances, they are sensitive to novel, man-made changes. For example, the photosynthesizing zooxanthellae require sunlight and so any pollution, even by inert particles, that produces cloudiness in the water is fatal; likewise the polyps are killed if their calcareous tubes are broken, as occurs when people walk on them.

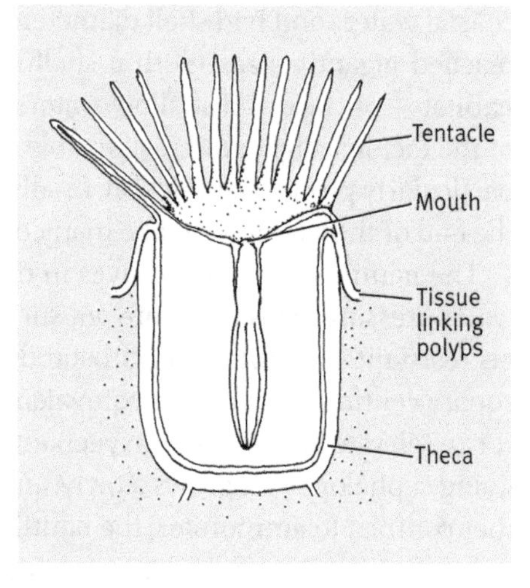

FIG. 8.10 Scleractinian coral polyp in sectional view

Time warp in amber

Conifers and some related trees produce resin that extrudes from the trunk or twigs if these are damaged. This is a defensive mechanism: an invading insect may be washed out and imprisoned and the resin seal prevents the entry of fungi into the tree. In time the resin will harden and eventually, when of the consistency of a soft rock, it will survive for millions of years and is known as amber. When newly extruded, the resin is very sticky and a variety of small organisms become trapped; if these are totally enclosed they will be protected from decay and provide uniquely well-preserved fossils in which the details of the living animal can be seen. Many insects closely related to present day species have been recovered from ambers as old as 60–55 Mya, but recently a study of amber of Triassic age has revealed a variety of micro-organisms. These included several different protists that could be assigned to modern groups, showing that in these single-celled organisms evolution in structure has been very slow; the community was not very different from one that might be trapped in resin today.

Triassic extinction

This extinction is really the least understood of the five major extinctions. Over the last 20 million years of the period large numbers of families of vertebrates and marine invertebrates disappeared from the fossil record and there were important regional extinctions amongst the plants. There is evidence to suggest that there were really two events, one in the age called the Carnian (227–220 Mya) and one at the end of the Triassic period (206 Mya). Michael Benton has pointed out that the former event seems to have been most significant for tetrapods, as thirteen families became extinct at that time and only six at the end of the period; among the reptiles these changes represented a shift from non-dinosaurs to dinosaurs. There were also many extinctions among the ammonites, the sea lilies and echinoids and the various reef builders; the scleractinian corals became the main framework builders only in the last 15 million years or so of the Triassic, but suffered some extinctions themselves at the end of the period. A most notable extinction at the end of the period occurred in the ammonites, of which it seems that very few genera survived. Numbers of families or genera of many other groups of marine invertebrates also became extinct and conodonts disappeared from the fossil record.

It has recently been suggested by the Canadian scientist John Spray and colleagues that a fragmented comet may have struck the earth around 214 Mya. They have linked the Manicougan crater in eastern Canada with one near St Martin in central Canada and with the Rochechouart crater in France. They also suggest that two smaller craters, in the USA and Ukraine, may have resulted from the same comet. A signature of asteroid or comet impact is normally an increase in the level of the rare metal iridium in the rock layer from the period. No iridium layer has been found in rocks from the last part of the Triassic, but some comets have a low iridium content. It has been noted by palaeontologists that the earlier Triassic extinction event was particularly marked in Laurasia (now North America and Europe) and this would fit with the distribution of this fragmented comet. However, the present dating of the extinction event, around 220 Mya, is not compatible with that for the comet.

The sea level fell markedly at the end of the Triassic and there is some evidence that sharp falls were followed by sharp rises. In such a situation anoxic waters flow over the bottom of shallow seas and such events would account for the loss of so many bottom-dwelling (benthic) organisms. They would not, however, explain the extinction of tetrapods and changes in the flora. There is clearly much more to be found out before the events at the end of this period are properly understood. But whatever happened was to herald the advent of a period with a largely benign environment which saw the evolution of large numbers of new and spectacular species.

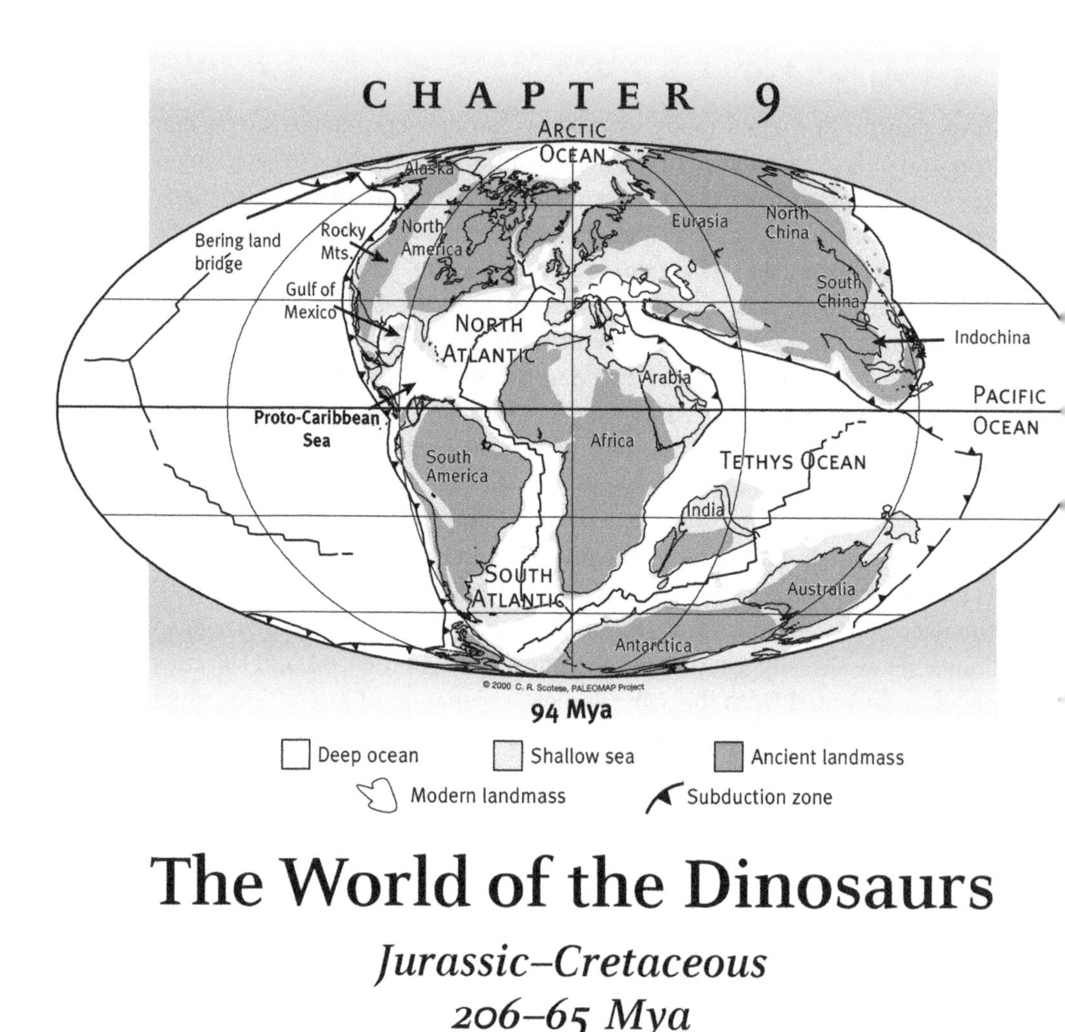

CHAPTER 9

The World of the Dinosaurs

Jurassic–Cretaceous 206–65 Mya

THROUGHOUT the Jurassic and Cretaceous periods the movements of tectonic plates were causing the giant continent Pangaea to split up, the process becoming particularly marked in the Cretaceous (145–65 Mya). During the Jurassic the central Atlantic opened and, linking with the Tethys Ocean, split Laurasia in the north from Gondwana in the south. By the Early Cretaceous strong westerly currents flowed through these oceans, between the large land masses. At the same time the southern parts of Gondwana were dividing—the South Atlantic was born as a narrow ocean. On the other side of Africa the Indian Ocean was beginning to form as the tectonic plate that carries India began its dramatic movement north. Africa was also moving north and the result of these two processes was that early in the Tertiary the Tethys Ocean was squeezed virtually out of existence. Throughout the period, until near the end of the Cretaceous, the sea level rose so that there were many shallow seas on the continents: it is estimated that around one third of the present land surface was covered with water. For example, North America was divided longitudinally by the Interior Seaway that ran from the Arctic in north Alaska through to the Gulf of Mexico. The fall in the sea level at the end of the period was caused by the land being pushed up to form great mountains—the Himalayas and the Alps—as the tectonic plates collided.

In the early stages of the break-up of Pangaea, some parts remained largely intact so that in the Jurassic many areas had an extreme continental climate with marked seasons and little rain; these arid conditions returned at the end of the Cretaceous when the sea level fell. From the Early to mid-Cretaceous the global temperature was high. Swamp forests flourished along the borders of the Tethys Ocean, and its waters contained an abundance of micro-organisms. As the leaves and trees of the forests fell, peat was formed on land; eventually—under compression and heat—it became coal, while the marine deposits gave rise to the majority of the oil that we exploit today.

Throughout the period the movement of the tectonic plates caused frequent episodes of volcanic activity. There was a particularly large eruption at the end of the Cretaceous when a great outpouring of basaltic rock occurred in India forming the Deccan Traps. There is evidence that at a number of times in the Jurassic and Cretaceous there were limited extinctions of the flora and fauna; at the end of the Cretaceous there was a massive extinction affecting life both on the land and in the sea.

Throughout most of the periods covered in this chapter the vegetation was dominated by gymnosperms—cycads, conifers, monkey puzzles, yellow woods, maiden-hair trees and related forms. They grew in a very wide range of habitats from arid mountain to the borders of seas (as mangroves do today). Ferns, clubmosses and some horsetails formed the shorter vegetation in many sites for much of the time, but in the Cretaceous there is the first evidence of the evolution of flowering plants (angiosperms). The fossil record suggests that they originally occurred in the lower latitudes (in or near the tropics) and particularly in disturbed sites alongside rivers. It is easy to envisage that the giant dinosaurs would have created many disturbed habitats as they moved around.

Towards the end of the Cretaceous the flowering plants became very diverse and widespread and their evolution will be reviewed later.

Types of dinosaur

The first dinosaurs appeared in the Triassic (see FIG. 8.7) and, prior to their demise at the end of the Cretaceous, they were the dominant terrestrial animals for over 150 million years. However, during this period many species evolved and many became extinct, five to ten million years being a typical 'life' of a species. Nothing like all the dinosaurs that occurred have been found; the total number of species was probably around a thousand, of which about a quarter are known at present. As the supercontinent was breaking up, most species had ranges that were restricted to a particular region. Therefore only a relatively small number of the dinosaur species that are known actually lived in the same countryside. Most were either on different land masses or lived in different periods.

Dinosaurs are divided into two major groups based on the structure of the hip (FIG. 9.1); they are known as the bird-hipped (ornithischian) and lizard hipped (saurischian) dinosaurs. The names are rather misleading as it is from the saurischians that the birds actually evolved; these names were given before this was fully understood. The pelvic girdle consists of three bones: the ilium attached to the backbone and the ischium and pubis which may crudely be described as holding-in the rear of the animal. In humans the position of the ischium means that it takes the weight when we sit, while the pubis is in the front. Together these three bones enclose the hip socket of the hind leg. In the saurischians the pubis points forward, but in the

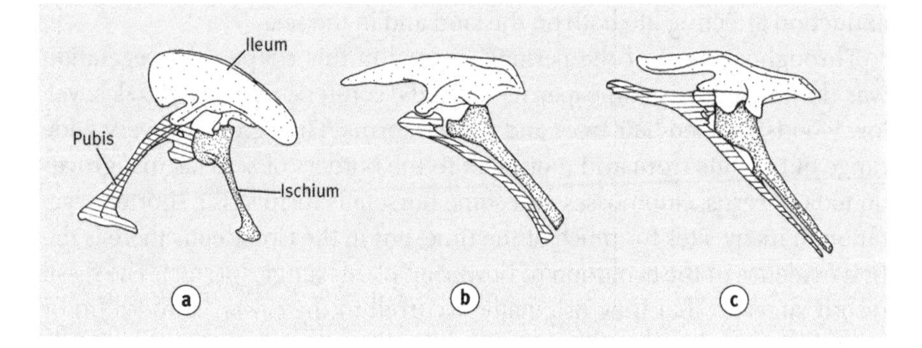

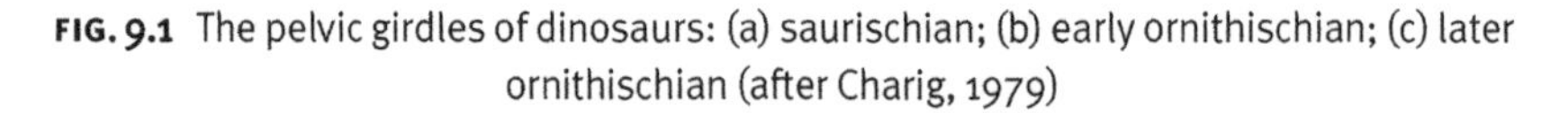
FIG. 9.1 The pelvic girdles of dinosaurs: (a) saurischian; (b) early ornithischian; (c) later ornithischian (after Charig, 1979)

ornithischians it points backwards alongside the ischium. This difference can be understood if we consider their feeding habits and stance when walking. All ornithschians were herbivores and many were bipedal, walking for at least part of the time erect on the hind legs. As vegetarians they would have a large gut to allow the food to pass through sufficiently slowly to allow it to be digested, a process involving symbiotic bacteria. We can imagine an erect ornithischian as having a 'beer belly' and if this had rested on a forward pointing pubis it would have overbalanced the animal. The backward pointing pubis permitted the gut to hang between the legs. The forward projection, found in the Cretaceous species, may have helped suspend the gut. The bipedal saurischians (*Tyrannosaurus* and relations) were all carnivores so their guts would have been much smaller as meat is quickly digested.

The ornithischians are usually divided into five groups: the ornithopods and pachycephalosaurs that were bipedal and three mainly quadrupedal groups all of which were to some extent armoured—these are the plated dinosaurs (stegosaurs), armoured dinosaurs (ankylosaurs) and the horned dinosaurs (ceratopsians). The earliest ornithopods, like *Heterodontosaurus* (see FIG. 8.7), occurred at the very end of the Triassic and early in the Jurassic; they were vegetarians. Some bore a pair of large tusks, which probably served a similar function to the tusks of wild pigs—defence and aggression between males.

Late in the Jurassic another group of ornithopods appears in the fossil record. These are the iguanodontids, the best known of which is *Iguanodon* (FIG. 9.2), one of the first dinosaur fossils to be discovered (in 1822). Members of the group occurred until the end of the Cretaceous, but by then they were limited to Europe. In the early and middle part of that period they were at their most diverse and widespread. The species ranged in size from five to ten metres; *Iguanodon* was one of the largest and probably weighed five and a half tons, equivalent to a large elephant. How did they stand? Although they would have been able to stand upright, to look out for enemies and to browse tall trees, the structure of the skeleton suggests that for much of the time, and especially when they ran, the backbone would have been in a horizontal position—only the back legs would have been used and the thick tail held well away from the ground, counterbalancing the front of the body. This posture would have been rather like that we can see in many birds (for example, road-runners, blackbirds or pheasants) when they run quickly. If the iguanodontids grazed on low vegetation, or if their large adults ambled, they may have used all four legs, the short forelegs support-

FIG. 9.2 Early Cretaceous landscape in south-east England. The fish-eating theropod *Baryonyx* and a group of small ornithopods, *Hypsilophodon*, are in the foreground and an armoured dinosaur, *Polacanthus*, and small herd of *Iguanodon* in the background

ing the front of the body, as happens with kangaroos today. It seems as if all bipedal ornithopods would have moved in this way.

Particularly abundant ornithopods were the duck-billed dinosaurs (hadrosaurs) that flourished towards the end of the Cretaceous. So far fossils have been found almost worldwide. Some of them had remarkable, partially hollow, horn-like extensions on the skull (see FIG. 9.5), and these animals are often referred to as the crested dinosaurs. The function of these structures has been the subject of much debate, but their role seems most likely to have been in signalling. They may have been brightly coloured which would both allow members of the same species to identify each other easily and also serve to establish a social hierarchy (or pecking order) within the herd; those with the most spectacular crests usually taking the top place. These crests are connected to the nasal passages and it is also probable that they amplified and modified snorts: reconstructions, both physical and virtual, have shown that when blown through they would have made a noise like a trombone. The base of the tail was especially deep, a good design for swimming, and this has led to the suggestion that they were aquatic. However, when a hadrosaur was found, surprisingly with its stomach content intact, this consisted of pine needles and other fibrous terrestrial vegetation. Thus, it is likely that duck-billed dinosaurs grazed around the margins of lakes, rivers, and shallow seas, but were able to take to the water if attacked by predators. Although covering the same range in length as the iguanodonts they seem to have been a little less robust and consequently weighed less; perhaps they ran faster for, unlike many of their contemporaries, their only defence seems to have been escape by running or swimming.

Apparently much less abundant than the duck-billed dinosaurs were a related group, the pachycephalosaurs or dome-headed dinosaurs (FIG. 9.3), that lived mostly in the Late Cretaceous. They had a greatly thickened skull roof that gives them their name. It is suggested that they lived in herds and that the thickening served to protect the brain when they butted a predator or rivals in the herd—much in the way that sheep, particularly rams, behave today.

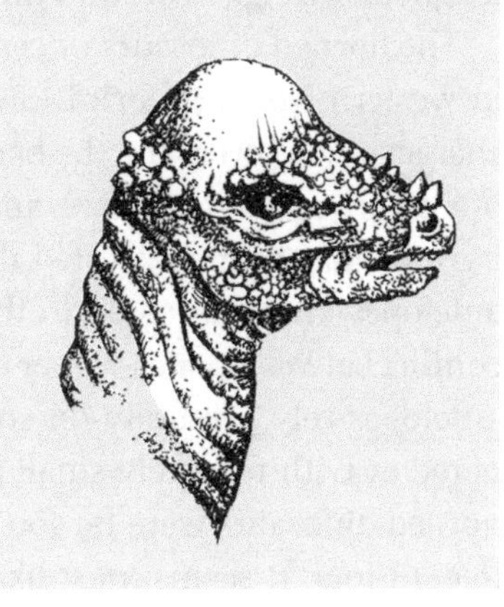

FIG. 9.3 Head of a pachycephalosaur (after Sibbick in Norman, 1985)

The other three groups of ornithischians were all armoured to some extent, often to a very great extent; they had stout bodies and walked relatively slowly on all four legs. The stegosaurians or plated dinosaurs (FIG. 9.4), are found from the mid-Jurassic onwards in many parts of the world. Their characteristic was a series of thin bony plates along the back, which have numerous small canals and were obviously well supplied with blood. It is thought that these will have served for heat transfer. First thing in the morning, positioned to face the sun they would have functioned like solar panels and warmed up the dinosaur; in the heat of the day, with the plates held so they were end-on to the sun, any wind would blow through them, so they would have acted like a car radiator and cooled the blood. Some have suggested that these plates may have served as an armour, but the presence of a blood supply argues against this; for defence, the plated dinosaurs would have depended on the formidable spikes near the end of the tail.

The armoured dinosaurs or ankylosaurs (FIG. 9.2), which lived in the latter part of the Jurassic and in the Cretaceous, were particularly diverse towards the end of the latter period. The bones in the skin were often fused together to form numerous thick shield-like pieces of armour, sometimes extending as broad spikes; the tail of most species found later in the Cretaceous was modified into a thick and hard club, and some even had bony eyelids. They have truly been described as the 'tanks' of the dinosaur world. Like the plated dinosaurs they ranged up to six or seven metres in length and weighed around three and a half tons.

The horned dinosaurs or ceratopsids (FIG. 9.5) were relatively abundant in western North America towards the end of the Cretaceous. They were characterized by having the back of the head extended into a thick frill, that often bore spikes or bosses, and in most species the head had one or more horns. It has been suggested that they should be compared with deer and antelopes, their horns and other armour being more for display and for conflict between males, rather than for defence. On the other hand, deer and antelopes rely primarily on speed to escape predators (though when cornered or with relatively small predators they will use their horns); but the horned dinosaurs were far too bulky, weighing up to six tons, to have been fleet of foot. It seems most likely that, as so often in evolution, more than one force has driven the adaptation, these structures having been used both for defence and for social combat and display.

FIG. 9.4 Late Jurassic landscape in North America. The plated dinosaur, *Stegosaurus*, and a group of small theropods, *Ornitholestes*, are in the foreground, while a herd of giant sauropods, *Diplodocus*, passes in the background

FIG. 9.5 Late Cretaceous landscape in western North America. A distant *Tyrannosaurus* threatens some horned dinosaurs, *Triceratops*, while three crested dinosaurs, *Parasaurolophus* flee from the scene and startle two small theropods, *Dromaeosaurus*, feeding on a young ornithopod

The Saurischia fall into two very dissimilar groups: the Sauropodomorpha—quadruped and often gigantic herbivores (FIG. 9.4)—and the Theropoda, bipedal carnivorous predators, ranging in size from under a metre to the terrifying 14 metre *Tyrannosaurus* (FIG. 9.5). Occupying the niche formerly held by the prosauropods (see FIG. 8.7) which, however, were not their ancestors, the sauropodomorphs first appeared in the latest Triassic; in the Jurassic they were particularly diverse and widespread. After this period they seem to be missing from the fossil record for western North America, where the hadrosaurs and ceratopsids were the abundant herbivores. However, recent discoveries in various parts of what was Gondwana, show that they continued to flourish there and in Europe well into the Cretaceous. One feature of the sauropods is their long neck and it was believed that this enabled them to browse the tops of tall trees. Some studies made in 1999, based on digital reconstructions of the necks of two well-known dinosaurs, *Apatosaurus* (formerly called *Brontosaurus*) and *Diplodocus*, threw doubt on the generality of that view. In these species it seems that the neutral position was with the neck straight and more or less horizontal, so that the heads which were angled down would be relatively close to the ground; the necks could not have been easily moved upwards, particularly in *Diplodocus*. If these ideas are correct it suggests that these animals may have browsed at bush height rather than tall trees. As I will describe later this suggestion can be supported on ecological grounds.

All sauropods were large animals, mostly gigantic. *Diplodocus* (FIG. 9.4) measured some 27 metres from nose to tail and weighed 18 tons—equilivant to three large elephants. But this was a relatively modest species; it is estimated that *Apatosaurus* would have weighed 28 tons and that it would have been dwarfed by *Brachiosaurus* estimated at 25 metres in length and weighing about 55 tons. It has been claimed that even larger dinosaurs lived in North Africa and South America, but fairly complete skeletons are yet to be found. Such large animals must have moved their limbs slowly, keeping their legs straight to avoid overloading the skeleton; but of course each stride would have been a long one so their speed relative to the ground could have been quite reasonable. The elephant also moves its legs in this way: it does not run or gallop, but when walking fast moves more quickly than a running man. Sauropods have been found on all present-day continents, other than Antarctica (most of which is covered by deep ice preventing fossil recovery). They did occur in south-east Australia which at that time was within the Antarctic circle; the global temperature was high in the

Cretaceous and there were no polar ice-caps, nevertheless the climate there was probably cool temperate and the position within the polar circle would have meant long periods of darkness in the winter months. This has raised the question: did these dinosaurs have night vision?

The theropods (FIGS 9.2, 9.4, and 9.5), found in the Jurassic and right to the end of the Cretaceous, were all predators feeding on other animals; these ranged from insects to early mammals and other dinosaurs, depending on their relative sizes. At least some hunted in packs. Compared with vegetarian dinosaurs they would have been relatively slim and correspondingly more fleet of foot. Probably all were fully bipedal, moving entirely on the hind legs which bore powerful claws that would have been used to tear at their prey. The front legs were short and also had claws. In some, such as *Tyrannosaurus*, they were so short we can only speculate on their function. Certainly they were not functionless for the shoulder muscles were very powerful. *Tyrannosaurus* was strongly built, reaching 14 metres and weighing perhaps 7.5 tons, and probably preyed on the slower herbivores such as *Triceratops*. *Albertosaurus* was more lightly built and so may have been able to catch the faster duck-billed dinosaurs. Large as *Tyrannosaurus* was, a theropod thought to have been even larger has been found in South America and called is *Giganotosaurus*. It has been suggested that *Tyrannosaurus*, and perhaps other large theropods, fed on carrion rather than capturing healthy prey; they may well have included carrion in their diet, but if they did not hunt living dinosaurs it is difficult to understand why ankylosaurs and ceratopsids evolved such elaborate armour (FIGS .9.2 and 9.5).

The lives of dinosaurs

At what pace did dinosaurs live—were they cold-blooded or warm-blooded? This question has been vigorously debated and the answer seems to be neither: they were intermediate with a system or systems unlike anything in living animals. Owing to their great size the body temperature of most species would not have mirrored that of the surrounding environment, except perhaps when they were quite young. Once such animals had warmed up they would take a long time to lose the heat, furthermore the gut of the large herbivores would have generated a lot of internal heat. Even modern cold-blooded reptiles are not entirely dependent on the surrounding temperature: various behavioural adaptations, like sun-basking, can raise their temperature significantly and allow them to move quickly.

However, there are pieces of evidence that suggest that dinosaurs did not maintain a constant temperature, for example their bones show lines of arrested growth such as are found today in reptiles that cease to grow in the winter. But another feature of the bone structure points the other way, the presence of what is termed 'dense Haversian bone' that is today only found in birds and mammals. This is formed when the bone is growing and layers are laid down and reabsorbed repeatedly so that the canals in the bone are particularly densely packed.

Some research workers have recently claimed that iron stains in the chests of fossil dinosaurs can be interpreted as a four-chambered heart with two separate circulations (like mammals and birds). This would have been needed in the largest dinosaurs where high pressures would be required to circulate blood to the head when the animal held it up since, if the body and lungs shared a common circulation, this pressure would have ruptured the lungs. Does this sophisticated heart mean that they were like mammals and birds and maintained a steady temperature independent of that of the air? If so, then the smaller species and young individuals would have had to have the type of insulation provided by fur or feathers; and although dinosaur skin is often fossilized, there is evidence for such coverings only in some theropods. Further evidence against a fully warm-blooded status in dinosaurs (including theropods) is provided by the apparent absence of turbinate bones in the nose; these act as sites for the condensation of water when we exhale, but they are delicate structures which may have existed but not have been preserved.

There are also arguments based on features of the dinosaur lifestyle. Several pieces of evidence suggest that young dinosaurs grew up quickly, which would depend on their maintaining a warm body temperature; the high global temperatures in this period would at least have assisted, but today in the tropics crocodiles take around ten years to become adult. Being so much larger one would have expected the giant *Apatosaurus* to take longer, but estimates, based on bone structure, give a similar time for its juvenile period—therefore they must, in some sense, have been warm-blooded. Some dinosaur fossils have been found at high latitudes where daylight hours would have been few and, even in the Cretaceous, winter temperatures would have been low. It is suggested that these species must either have been warm-blooded or they would have had to excavate burrows in which to hibernate or migrate to warmer regions. As yet, none has been found with feet adapted for burrowing, but migration may not have posed a problem for the large species. Measurements made between successive footprints on dinosaur trackways give stride lengths of between two and four metres, so even if they moved their legs slowly they would have covered quite a lot of ground. These trackways also show that many were walking in the same direction, supporting the idea of migration. However, uncertainty remains as to the behaviour of the young: did they remain in warm climates until mature? It seems quite probable that different groups had different thermoregulation regimes and the young may have differed from the adults. Certainly dinosaurs were not entirely like either modern reptiles (or birds) or modern mammals: in respect of their body temperature, as in much else, the dinosaurs were unique.

Many dinosaurs were social animals: not only do the trackways show numbers travelling together, but in some places great numbers of fossils have been found showing that an entire herd had perished in some calamity. The most spectacular example is of an estimated 10,000 duck-billed dinosaurs. Judging by modern mammals and birds there would have been a social structure within the herd: some animals would have been dominant, others low in the pecking order. As has been pointed out both the crested duck-billed and dome-headed dinosaurs had head ornamentation which is best interpreted as having a social function. Modern equivalents may be the bleating of a flock of sheep, and the thick skulls and horns of rams. The head shields and horns of ceratopsids probably played a part in their social life as well as in defence; collections of skeletons confirm that they were herd animals. Computer simulations show that the giant sauropods could have moved their tails at supersonic speeds which would have created a sound like a bull-whip; these sounds may have been the means of communication within the herd, like the honking of geese in a flock.

Social life extended to family life. Some dinosaurs laid their eggs in nests, and a fossil of *Oviraptor* has been found showing that the female died in the act of tending the nest. Sites have also been found where the nests clearly were made year after year in the same place and were often packed close together like those of many seabirds. Sometimes the parent or parents remained with the newly-hatched young and in some species it appears that the juveniles were also around the nest; as happens today with certain birds and mammals, they may have helped in the care of their young siblings.

It does not take much imagination to visualize the destruction wrought on a patch of vegetation by a herd of dinosaurs. Today in Africa one can witness the impact of a herd of elephants which typically weigh four to six tons; many dinosaurs were at least a little larger, whilst the sauropods were often four to ten times as large. Elephants sometimes uproot trees by standing up on their hind legs and pushing, and the dinosaurs may well have done the same; but the trampling of plants and bushes is likely to have been the major effect. After ground is churned up in this way it is colonized by fast-growing 'weedy' plants. These plants would have grown at a faster rate than the damaged trees and shrubs.

These great assemblies of dinosaurs would have needed a lot of food. It has been estimated that the weight of dinosaurs per square kilometre on the Morrison Plain, Utah, in the Late Jurassic (see FIG. 9.4) could have been nearly twenty times more than that of the mammals in one of the crowded plains in Africa today. How did they get enough food? We must assume that they did not need as much and this supposition is supported by the fact that the giant tortoises on Aldabra Island in the Seychelles exist at densities where the weight per square kilometre is slightly over half that estimated for the dinosaurs. (Incidentally, this is another piece of evidence for herbivorous dinosaurs not being fully warm-blooded—they would have needed too much food.) But the dinosaurs would have been able to get more food if they thoroughly disturbed the ground and fed several months later on the fast-growing plants that colonized these places. This is why it is interesting that those making computer models of sauropod necks have suggested that some of them could not easily raise their heads to their full height, but held their necks horizontal with the head towards the ground. Rather than wandering in park-like landscapes, should we visualize dinosaurs on fern praries or amongst thickets of shrubs with the growth features of willows? These grow so quickly that they are used today as 'biofuels' and they withstand trampling as every shoot pushed into the soil forms roots. Towards the end of the Cretaceous, the flowering plants evolved, apparently in disturbed habitats. Did the habits of the dinosaurs provide the opening for the evolution of the plants that dominate our world today?

The more traditional view of the feeding habits of the sauropods is that they were like giraffes, holding their heads high as they browsed trees, up to 14 metres high in the case of *Brachiosaurus*. The very build of this dinosaur with its particularly long front legs looks like an adaptation to browsing. The teeth of *Diplodocus* seem to be ideal for stripping the foliage from branches in a raking movement, a process that would have been assisted by some kind of tongue which it is thought they had. From studies of the teeth and the way they had been worn, it seems that other sauropods sliced though twigs or small branches. The leaves were not chewed in the mouth, but quickly swallowed. These dinosaurs had a gastric mill with stones (gastroliths) that would grind up the food prior to its slow progress through the vast gut, in which we may suppose there would have been active microbial symbionts. The trees on which they would have fed were gymnosperms, related to today's monkey puzzles, redwoods, firs, and pines. They have particularly tough, spiky, and resinous leaves and twigs, which few modern animals find palatable. So to feed on them the dinosaurs must have had particularly tough mouths and throats and a 'good' digestive system. Many of these trees grow to great heights and when their lower branches are removed they are seldom regrown. One can speculate that if these giant sauropods browsed on forest composed of these trees it would be a long time before they would again find enough food there. Perhaps it was this, rather than climate, that led to the migrations—evidence for which is preserved in the trackways.

It is considered that the various armoured and many of the duck-billed dinosaurs fed on low vegetation; the spread of flowering plants in the Late Cretaceous period may have had a direct association with the evolution of the diverse and often abundant dinosaurs of that period. The beak-like extension of the jaw in the duck-bills would have been well suited to gathering plant material, while the jaw itself was hinged in a way that allowed them to chew and so break up the food. This would have aided digestion and eliminated the need for the enormous gut of the sauropods.

All these herbivores would have been potential prey for the carnivorous theropods. Adult sauropods were probably protected by their size and may have lived to a great age; they also laid many eggs at a time. These two facts suggest that in their youthful stages they must have suffered heavy mortality—otherwise the population would have grown ever larger. We see a similar pattern, of high mortality in the young and long survival of the adults, in giant turtles and crocodiles today. Young ankylosaurs and ceratopsids would have gained some protection from their armour, but like young rhinoceroses would have been vulnerable to many enemies that posed no threat to their parents. Perhaps the ceratopsids lived in herds of mixed ages, the larger adults forming a protective ring as do musk oxen against wolves. The ornithopods' only defence in close combat, apart from size, was probably the strong clawed thumb on the front limbs of a few species (see FIG. 9.2). Fleeing would seem to have been the best tactic and the bipedal stance would allow them to keep a good look-out for enemies. They were more heavily built than comparable predators so they would have needed to have had an advantageous start if they were not to be overtaken; some may have escaped by swimming. It is likely that in common with modern reptiles any chase could not be prolonged as the oxygen supply to the muscles would soon have been exhausted.

There is evidence that some carnivorous dinosaurs hunted in packs, like wolves, which would have enabled them to overcome larger prey. However, judging by the prowess of the Komodo Dragon of Indonesia—a giant lizard, up to three metres long, which can kill a buffalo—these theropods may have been able to take prey as large or larger than themselves. Studies of the head suggest that in some species this was strong enough to withstand the shock if the dinosaur attacked its prey by running at it with open jaws. The powerful claws on the hind legs are likely to have been used to disembowel the victim. In some theropods the front legs could have served to aid the attack, but in others like *Tyrannosaurus* and its relations, the role of the very short front leg remains, like much about the dinosaurs, a fascinating enigma.

Marine monsters

Large air-breathing reptiles flourished in the seas as well as on land. The initial return to a marine lifestyle occurred in the Triassic, but it was in the periods covered in this chapter that they achieved their greatest diversity.

The ichthyosaurs, with their streamlined tuna-like body (FIG. 9. 6), were the porpoises of the Jurassic. The group was very diverse in the Triassic; the species that lived then were less streamlined than the later forms, but some reached a considerable size. The largest was *Shonisaurus* which measured some 23 metres in length, with a robust and very deep body that could have contained a substantial gut, and a pointed head with small eyes and teeth. Being air-breathing, ichthyosaurs would have tended to be buoyant, in contrast to sharks, with which they are sometimes erroneously compared. Like porpoises they would have had to squeeze air out of their lungs to dive and hunt fish, cephalopod (such as squid), and other prey. We know their diet because some fossils have been found with the remains of their last meal in the body. Even more remarkable fossils are those where the ichthyosaur died in the process of or just before giving birth. The young was expelled tail first, and the female would have had to bring it to the surface quickly and support it while it took its first breaths; this happens with whales. Although apparently abundant in the Jurassic, ichthyosaurs became rare in the Cretaceous, apparently dying out before the end of the period.

FIG. 9.6 An ichthyosaur (after Charig, 1979)

Whereas we can visualize ichthyosaurs following a very similar lifestyle to that of modern porpoises, those plesiosaurs with long necks and small heads were quite unlike any living animal, unless of course the 'Loch Ness Monster' turns out to be real! (FIG. 9.7). The four legs were modified to form strong paddles, which probably functioned like those of turtles, though some other studies suggest they could not be raised above the horizontal. They had very sharp teeth showing that they fed on fish and other marine animals, which they may have caught by darting their head forward on the long neck. One can imagine them hunting along coasts and grabbing a large crab or similar victim before it had time to flee for cover in a nearby rock crevice. There were other plesiosaurs that had necks of more normal lengths and large heads, these were the pliosaurs. Some were of a remarkable size; *Kronosaurus* reached 12 metres in length and would, like the killer whales today, have been a top predator in the oceans of this epoch. But some pliosaurs took small prey, and a fossil has been found with the remains of many small ammonites in its stomach. Plesiosaurs first appeared in the Late Triassic, but became extinct at the end of the Cretaceous.

The mosasaurs were large marine lizards (4.5 to 9 metres long). Their limbs were adapted for swimming but, as with modern lizards that swim, most of the propulsion was probably provided by the tail. A few had rounded teeth suitable for crushing shellfish (as seen in some sharks FIG. 7.14 (b) and (c)), but most had sharp, pointed teeth and, like the pliosaurs, would have been ferocious predators. They had only a relatively brief appearance

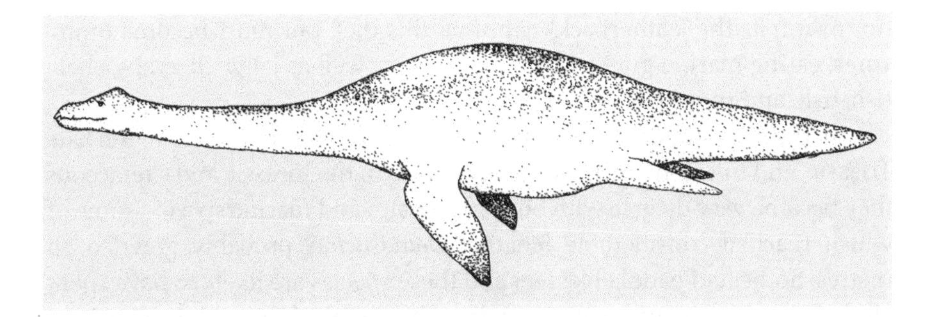

FIG. 9.7 A plesiosaur

in the kaleidoscope of life: having evolved in the Late Cretaceous, they became extinct at the end of that period.

There were two groups of reptiles in the Jurassic and Cretaceous seas that survived the extinction at the end of the period and which have members living today: the turtles and crocodiles. Some turtles occurred in the Triassic, but they became particularly diverse and large in the Jurassic and Cretaceous. One, *Archelon*, had a shell over three metres in length and flippers that spanned four metres. The shell, such an effective defence for terrestrial tortoises as they can withdraw completely into its shelter, is less valuable for sea turtles as they cannot entirely withdraw the head and flippers. Their main defence when full-grown must be their bulky size and, although adult turtles can fall victim to large sharks, it is the young that suffer prodigious mortality. We can suppose that these ancient turtles lived similar lives to their present day descendants, which remarkably navigate over great distances from many parts of the ocean to come, more or less simultaneously, to particular beaches where they lay their eggs. From 50 to over 100 eggs (depending on the species) are laid in a hole or 'pit' scraped out by the female in the sand. Eggs may be destroyed by other females digging them up when making their own nest holes. A common hazard is posed by enemies, such as monitor lizards, raccoons, foxes and pigs, which sniff out the nests and eat the eggs. When the young hatch they make their way down the beach to the sea with all possible speed, for a wide variety of predators will consume them. It is believed that the habit of synchronous laying and hatching was evolved to 'swamp' these enemies; some baby turtles will escape due to the sheer weight of their numbers. They then swim away from the beach of their birth, the few that survive returning many years later to lay their eggs. All sea turtles start off life as carnivores, some (for example, the leatherback) continue this diet, but most become omnivores, eating marine grasses and seaweeds, as well as jellyfish, crabs, shellfish, fish, and sponges.

Like the dinosaurs, the crocodiles evolved from the archosaurs in the Late Triassic, and the early forms were terrestrial. In the Jurassic and Cretaceous they became very diverse with both freshwater and marine species, some of which reached considerable lengths: *Deinosuchus* probably grew to 12 metres. Some had paddle-like feet and these species are likely to have spent all their lives in the water, except when the female dragged herself ashore for egg-laying; but the majority were probably amphibious as they are today. In their anatomy modern crocodiles are not unlike some of the ancestral forms and therefore information on their lifestyle may give us some insight into that of ancient species.

In spite of their predatory habits the teeth and jaws of crocodiles, alligators and their relations are not adapted for cutting; therefore, if they take large prey they have to tear off portions. They do this by taking a grip with their jaws and twisting, rotating their whole body; the neck is particularly strong to provide the force for the action of the jaws and to withstand the strain. Crocodiles and alligators feed on fish and air-breathing animals. The latter (reptiles, birds, and mammals) they frequently kill by dragging them under the water. They digest slowly, having strong hydrochloric acid, but few enzymes in their digestive juices. Therefore large prey are not consumed all at once but often stored, for example, under a river bank. This habit leads to their consuming carrion, which they will also take directly if the opportunity arises; their strongly acidic gastric juice provides protection against food poisoning. They spend much of their time resting on the banks of rivers and estuaries and they have a low demand for food, probably consuming little more than twice their own weight in a year. (This contrasts with small birds that often require their own weight of food in a day, a consequence of their being warm-blooded and of small size—which results in increased heat loss due to a large surface area in relation to their weight).

Female crocodiles that live in shaded areas gather a pile of vegetation and lay their eggs in this mound which ferments and provides heat to incubate the eggs. Those that live in open habitats dig a hole in the ground and lay around 50 eggs that are incubated by the heat of the sun. The mother stays near, often does not eat and will defend the site, but apparently she may doze off so that large lizards and others are able to dig out the nests; once opened the mother makes no attempt to rebury the eggs and all will die.

When the young are ready to hatch, after about two to three months, they make a croaking sound and the mother then uses her body to push away the soil down to the first layer of eggs. After hatching the young are taken down to the water, sometimes in the mother's mouth; she will often remain with them for a short while and then they are left on their own. Young crocodiles are ferocious predators, but their small size makes them vulnerable to many enemies, including larger members of their own species. They spend the early years of their lives away from the adults' habitat, amongst lush vegetation and in small water bodies. They become mature in about ten years, but only a few live to that age, though once adult they have few enemies apart from man and probably live for at least another 30 years. One can gain an idea of the level of mortality of the young from the simple calculation that laying an average of 50 eggs a year during her lifetime a female Nile crocodile would have produced 1500 eggs, but for the population to remain more or less stable, as it does, only two of these eggs would need to survive to adulthood.

A rich brew in the seas

The marine reptiles of the Jurassic and Cretaceous were at or near the top of the food chain; and the fact that they, and some sharks, were so large suggests that the start of this chain was very productive; that is there would be a high density of small organisms growing and reproducing rapidly. This is indeed the case; it is probable that the seas were more densely populated by photosynthesizing plankton than at any other period in geological history. They were fed on by other protists, by numerous filter-feeding shellfish (molluscs), barnacles, polyps, and worms. There were many ammonites and other cephalopods, shrimps, and early lobsters to take larger swimming prey, living or dead. Predatory molluscs, which bored holes in the shells of their victims, were more abundant than at any earlier time; also, in or moving on the seabed, were starfish, sand dollars, and sea urchins. The group of bony fish were diverse and numerous; among the cartilaginous fish the rays appeared for the first time. Over half the deposits of oil now being exploited were laid down from this rich brew in the Cretaceous.

Probably the most abundant photosynthesizing organisms supporting this great pyramid of life were the haptomonads, golden mobile protists found in the plankton. Most of these have a resting stage, the coccolithophorid (FIG. 9.8(a))—given a different name, for these two stages were long thought to be different organisms. When the free-swimming stage begins to change to the resting stage, calcareous plates are formed inside the cell which gradually move to the exterior. These plates are termed coccoliths and each one is little more than a small fraction of a millimetre in diameter, but it is the accumulation of such extremely minute, and beautiful, plates that have formed the white cliffs of Dover and the deposits of chalk in many parts of the world. One can hardly comprehend the astronomical number of organisms involved. The group is first found sparsely in the Late Triassic deposits and reached its peak towards the end of the Cretaceous. Very few species survived the extinction event at the end of that period, although new species have evolved since.

Another group of protists, the diatoms, were also abundant in the Cretaceous seas and oceans (FIG. 9.8(b)). They are primary producers with green-brown chloroplasts and shells made of silica. Today they are widely distributed in freshwaters and in the seas, including those in polar regions. The oldest known diatoms come from the Late Jurassic, and by the end of the Cretaceous they had become diverse and abundant.

Higher in the food chain than the groups just mentioned were the forams, or more correctly foraminifera, that also contributed to the sediments which formed the rocks from this period. They have a long geological history being known from the Cambrian and, as mentioned earlier, were abundant in the Carboniferous (see FIG. 7.17). They were very diverse in the Cretaceous, but were greatly affected by the terminal extinction event. The shells or tests of forams have a basic organic framework to which calcium carbonate and/or other material is added (FIG. 9.8(c),(d)). There is an aperture in one part of the test through which long thin extensions of the cell are extruded; outside the test these form a network in which they trap their prey—cyanobacteria, other protists, and even small eel worms and crustaceans. Many species live on the bottom of the sea, some in the oceans at depths up to more than 7000 metres. Others are planktonic. One group of these that first became common in the Middle Cretaceous and still flourishes today is the Globigerina. Their dead sand-grain sized tests sink and for aeons these have formed extensive 'globigerina oozes' over many parts of the sea floor. Most forams are just visible to the naked eye but some large species, ranging up to 15 centimetres in diameter, occurred in the Cretaceous and in the next period, the Tertiary, and these were probably the largest free-living single cells ever. (Reptile and bird eggs are single cells, but they are not truly free living.) Today's large forams (up to about three cen-

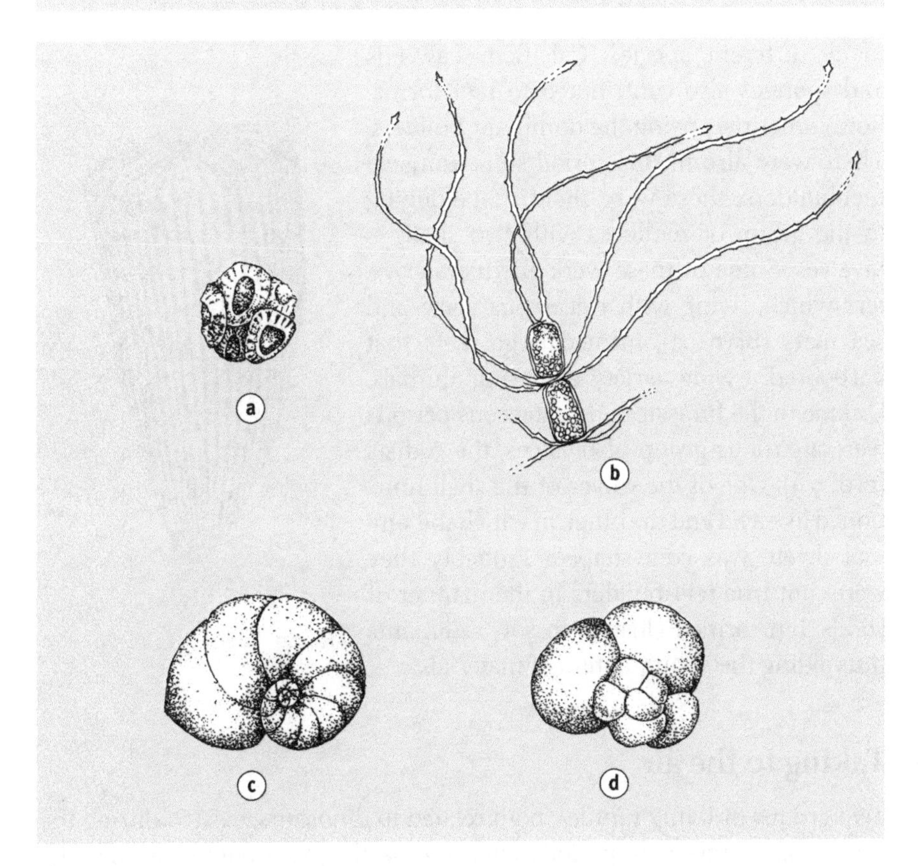

FIG. 9.8 Microscopic marine plankton: (a) coccolith; (b) diatom, *Chaetoceros*; (c) and (d) foraminiferan
((b) after Fogg 1968; (c), (d) after Parker & Haswell, 1898, from Bütschli)

timetres in diameter) live in well-lit tropical shallows and harbour symbiotic algae from which they obtain a large part of their nutrition; it is likely that the extinct species lived in a similar way. One can draw a parallel with the giant clams, which may have shells over half a metre long, live in shallows on tropical reefs, and have algal symbionts. It seems that this lifestyle, coupled with a strong shell, frees animals from some size constraints.

Just as the free-swimming life in the surface waters of the seas and oceans seems to have built up to a crescendo of diversity and abundance by the end of the Cretaceous, so did reef life. In the early Jurassic, reefs were rare but they started to become more widespread in the middle of that period, when the framework forming scleractinian corals (see FIG. 8.10) radiated into many different species. Calcified seaweeds and sponges also contributed to reef formation, sometimes being the dominant builders. There were also in this period some unusual reef-builders; these were shellfish, belonging to the group of molluscs with two shells—bivalves. Some of these were encrusting oysters which, living with calcareous algae and sea mats (Bryozoa), formed large reefs that harboured a wide variety of boring animals. Unique to the Jurassic and Cretaceous periods were a curious group of bivalves, the rudists (FIG. 9.9). One of the valves of the shell functioned like a lid and the other, in which the animal dwelt, was cone-shaped. Probably they were not true reef builders in the manner of corals, but formed clusters in soft sediments thus aiding the accumulation of material.

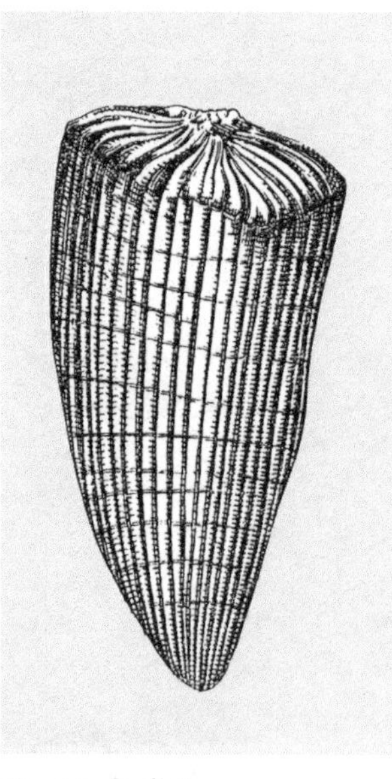

FIG. 9.9 Rudist

Taking to the air

Two groups of flying reptiles, both related to dinosaurs, existed during the Jurassic and Cretaceous periods. One of these, the pterosaurs, evolved in the Late Triassic and they were the first vertebrates to take to the air. They are widely present in the fossil record of this period and one species had the greatest wing-span of any known animal. The other group, birds, are not found in the fossil record until the Jurassic and there are not many fossils at all. In this period the birds would have seemed to have had a much less promising start than the pterosaurs, yet the pterosaurs became extinct at the end of the Cretaceous and in the Tertiary the birds radiated further and flourished. Once again we see that animals that are inconspicuous and apparently unpromising in one period prosper in the next, after an extinction has shaken the kaleidoscope of life.

Pterosaur bones and wings were fragile, light for flying, and therefore most of the numerous fossils that have been found are imperfect: the exact anatomy remains uncertain and a matter on which experts have contrary views. This applies particularly to the hind legs and their relationship to the hind margin of the wing. The front margin was supported by the very long fourth finger of the hand and it appears that the wing membrane itself contained some strengthening strands. Some years ago studies were made that suggested that the wing was narrow, independent of the hind legs and this scenario was extended further to the conclusion that pterosaurs could run, rather like birds, on their hind legs (that is, they were bipedal). Work on fossils recently discovered in Mongolia throws doubt on these ideas and suggest a more 'bat-like' animal, with the hind margin of the wing being attached to the hind leg. According to this interpretation they would have walked on all fours with the palms flat on the ground (FIG. 9.10), the latter view being supported by the arrangement of footprints in some fossilized tracks. These Mongolian fossils also showed that at least these smaller pterosaurs were covered with fur and so were presumably warm-blooded. Indeed it had already been pointed out almost a hundred years ago, that the speed and extent of muscular coordination necessary for flight would require an elevated and constant body temperature.

What was the start of the evolutionary path that led to flight in pterosaurs? There are three possibilities: they glided down from trees, they took off after running along the ground, or they leapt between branches on trees. If they walked on the ground as shown in FIG. 9.10(b), they would not have been able to run quickly or easily take off again after parachuting down from a tree. The most persuasive idea is that they jumped from branch to

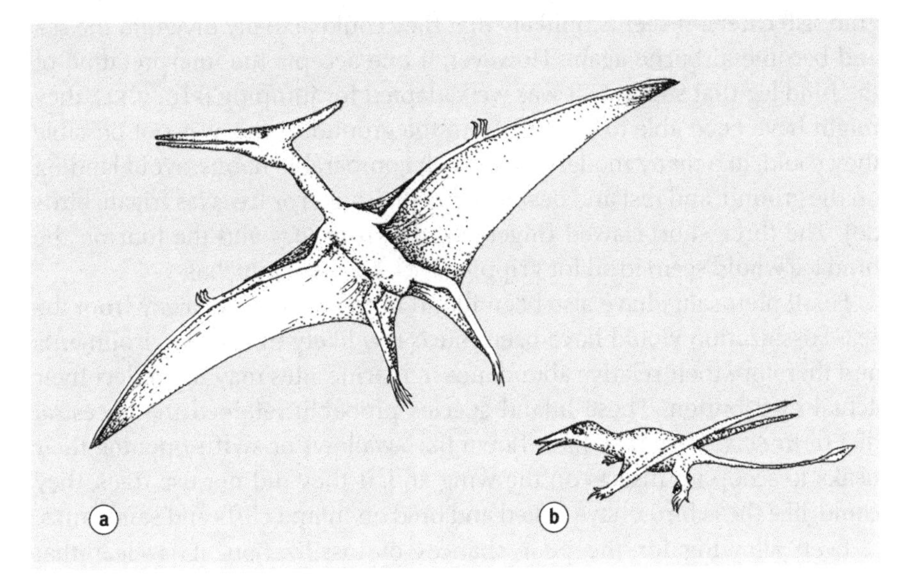

FIG. 9.10 Pterosaurs: (a) a pterodactyloid, *Pteranodon*, flying; (b) an early pterosaur resting on the ground

branch in trees, probably catching insects, so any growth of a flap along the side of the body between the legs would lengthen their jump and aid their landing, as such flaps do in modern flying squirrels. The earliest pterosaurs had long tails and teeth, but in the Late Jurassic and Early Cretaceous they seem to have been replaced by other species which had very short tails (FIG. 9.10(a)), and some of which were toothless. The latter group are termed pterodactyloids, though this term is sometimes erroneously applied to all pterosaurs.

Various suggestions have been made as to the food of pterosaurs. Perhaps the species with enormous wing spans, like *Quetzalcoatlus* (15 metres), flew like vultures scanning the ground for dead dinosaurs on which they fed. However their jaws and necks do not seem constructed for this rather tough task. It seems more likely that many were fish-eaters, such as the giant *Pteranodon* (FIG. 9.10(a)), with a 12 metre wing span, gliding almost effortlessly across the oceans and scooping up small prey, including fish, from the surface waters as the albatross does today. Here then were other 'monsters' depending ultimately on the rich brew of plankton in the Cretaceous oceans. The evidence from one fossil hints that *Pteranodon* may have had a pouch like a pelican; if so, the curious bony crest can be interpreted as a counter-balancing keel when the head was turned. The fossils of many smaller (pigeon-sized) pterosaurs have been found in rocks laid down in the sea or in coastal waters and they too are most likely to have been snatch and grab fish eaters. It seems unlikely that they could actually dive into the sea and become airborne again. However, if one accepts the interpretation of the hind leg that suggests it was well adapted for jumping (FIG. 9.11), they might have been able to take off from the ground. If this was not possible they could, like many modern birds with comparable habits, avoid landing on the ground, and rest and nest on cliffs (like auks) or trees (as frigate birds do). The three short-clawed fingers on the front legs and the four on the hind leg would seem ideal for gripping rock faces or branches.

Fossil pterosaurs have also been found in terrestrial sites away from the sea. Fossilization would have been much less likely in such environments and therefore their relative abundance in marine sites may not reflect their actual distribution. These inland species probably retained the ancestral diet of insects. They may have flown like swallows or swifts opening their beaks to scoop up insects on the wing and, if they did not use trees, they could, like these birds, have rested and bred on inland cliffs and sandbanks.

Even allowing for the poor chances of fossilization, it is clear that pterosaurs did not evolve to occupy as many roles as birds. As recent evidence suggests they had the front and hind legs linked by the wings so, like the condition in bats, they were literally tied together. By contrast, in birds, the two pairs could evolve independently, the front for flight and the rear for walking, swimming, climbing, prey capture, and a variety of other functions. The structure that permitted this separation was the feather—it maintains the rigidity of the wing through the tubular structure of the quill. In spite of its tremendous evolutionary significance for flight it seems that the feather came before flight, having been evolved for another purpose. This is yet another example of a structure with one rather minor function proving a key adaptation for a major evolutionary step (for example, fins evolved as feet for pushing through water-weed, coming to serve as limbs for walking on land, and insect larval gills evolved to pads for skimming across the water surface coming to serve as wings for flight) (pp. 73 and 109).

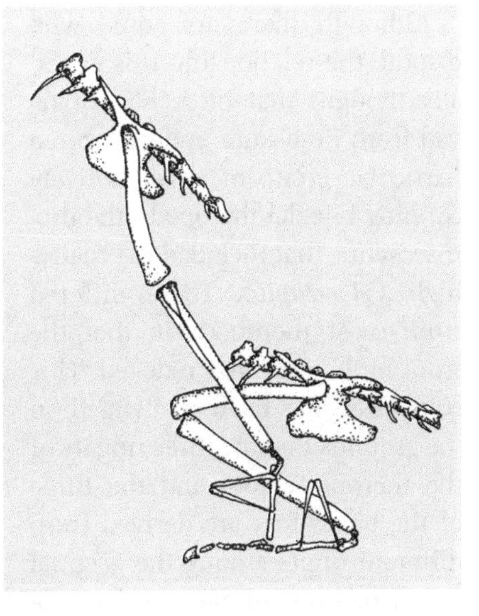

FIG. 9.11 Skeleton of pterosaur hind limbs when resting and when leaping into flight (after Bennett, 1997; in *Historical Biology* (www.tandf.co.uk/journals) by permission of Taylor and Francis Ltd.)

Various fossils of theropod dinosaurs, particularly those found in the Yixian Formation in China, show that they were covered with down and feathers. These rocks date from the Early Cretaceous, some time after the famous Late Jurassic bird, *Archaeopteryx* (FIG. 9.12). The fact that feathers have not yet been found for certain on earlier theropods makes interpretation difficult. These feathered dinosaurs come from a number of different groups within the theropods and therefore it seems that feathers were widespread, serving to retain heat and/or having a role in sexual display. If thermoregulation was their function then it may be that only small species and the young of large ones were feathered. As this covering occurred in so many different theropods it is probable that it first arose early in the Jurassic (when they all had a common ancestor) and only subsequently evolved its more complex role in flight.

Although there are some who dispute the relationship it is generally thought that birds have evolved from dinosaurs: indeed from a particular group of small, actively running bipedal theropods, the dromaeosaurs, that included the metre-high *Velociraptor*. These differed from most theropods in that the front limbs were not reduced. This conclusion has been challenged on the grounds that the three fingers of the theropod's foot and the three of the bird's foot are derived from different digits among the original five. However, studies on development have shown that in birds these are gene-controlled switches and are not of such fundamental importance. Because birds have evolved from within the dinosaurs they have all the latter's basic anatomical features and logically this leads to classifying birds as dinosaurs. However, the distinction between the two is useful in practical terms, though when we see a bird running quickly with its neck stretched out in front and its tail parallel to the ground we probably get as close as we can to seeing a long-extinct feathered dinosaur on the move.

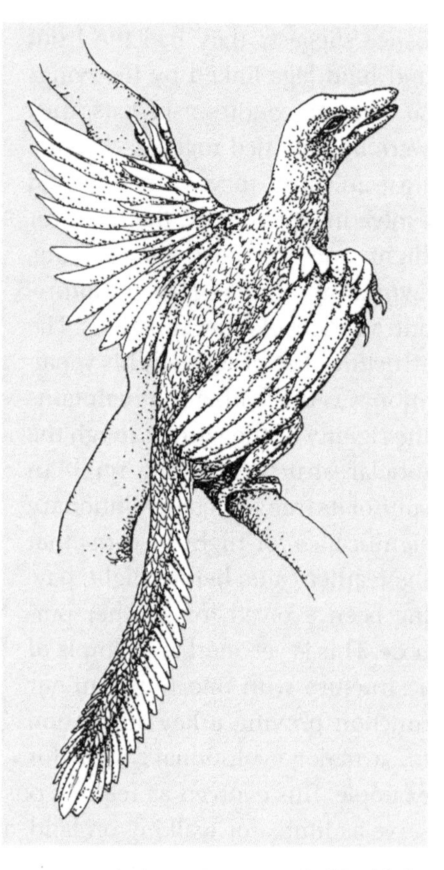

FIG. 9.12 *Archaeopteryx*—a primitive bird

How did flight first evolve in these reptiles? For many years there have been two competing theories, known as 'trees down' and 'ground up'. In the former it was proposed that living in trees the protobird glided between branches, increasing the length of its flight and improving its landing by flapping. However, the ground-up theory argues that if the protobird was a bipedal, running theropod it is more reasonable to consider that it took off after a run. The problem with this idea is that rudimentary wings would slow down running; sophisticated wings and considerable power are required to take off with flapping flight from a running position. How could each step in the intermediate stages in the evolution of the wing have been an advantage? A solution to this conundrum has recently been proposed: the protobird was an ambush predator jumping down from trees or rocks on its prey. The development of simple wings would improve control and manoeuvrability during the attack. A great virtue of this theory is that the various stages in the transformation of a forelimb to a wing can be traced through the evolutionary tree of theropods and birds; each new development occurs in the appropriate order to correspond with the ancestry of the animals; and each of these steps would give its possessor an advantage.

Studies based on the molecular clock (see FIG. 2.1) suggest that several groups of birds had evolved in the Cretaceous, indeed it is likely that *Archaeopterix*, with its many primitive features (claws on the front of the wing, toothed bill and skeleton in the tail) was something of a living fossil by the Late Jurassic. In subsequent periods birds became a major element in the fauna; today there are some 9000 known species, but it has been estimated that a total of 150,000 different species may have existed at one time or another. This diversity of birds is linked, like the much greater diversity of insects (FIG. 6.6), to the diversification of flowering plants with their particular flowers, fruits, and ecospace.

The promise of colour in the canvas of life

The ferns, conifers and other plants that had hitherto clothed the earth were predominantly green with a touch of brown, but in the middle of the Cretaceous the first flowering plants appear in the fossil record and after a while there were probably white flowers and later the rich palette of colours that we see today. This development of colour was driven by the coevolution of plants with animals, on which they depended for movement. There are two processes in the life of a plant where movement is an advantage, if not virtually a necessity: for fertilization, to bring the pollen to the ovule; and for dispersal of the seed away from the parent. Earlier plants (e.g. conifers) and some flowering plants, notably grasses and trees, such as oak and birch, rely on the wind to distribute pollen. Cross-fertilization (by the pollen from another plant) is likely only when large numbers of plants of the same sort grow together. This method requires great quantities of pollen (which is a curse for those of us who are allergic and hence suffer hay fever). In contrast a small amount of pollen carried by an animal, and most commonly an insect, can be targeted to another flower of the same species some distance away, especially if the insect is faithful to that type of plant. Thus we have the evolution in plants of many, often remarkable, adaptations that both make it advantageous for the insect to become a specialist and also discourage insects that are not specialists. These include specially shaped flowers, like a sweet pea that can only be forced open by a bee, or flowers with their nectar at the bottom of a long tube, which can only be reached by a moth (or humming bird) with a long tongue.

Although many insects eat the pollen or carry it to their young as bees do, the flower evolves to reduce this 'waste' and reward the pollinator with sugary nectar which is 'cheap' to produce, sugars being the product of photosynthesis. Nectar flows may be at particular times of the day which correspond with the flight time of their pollinator. There are complementary evolutionary forces that drive the insect towards specialization, so that it can be more efficient to feed at one particular type of flower and the insect can be more confident of success if this flower is, in a sense, reserved for it. Hence this tighter and tighter interlocking adaptation is described as coevolution.

A characteristic of flowering plants (angiosperms) is that the seed is enclosed in the wall of the ovary and sometimes other parts of the flower, to form a fruit. This fruit and the seeds in it, provide reserves of food so that when germination occurs the seedling can get off to a good start. The bigger the reserve the heavier the fruit or seed. Fern spores, which have very small reserves, can be blown hundreds of miles, but wind dispersal has a limited potential for flowering plants. Some have evolved special structures so that they can catch the wind—the plumes of the dandelion and the wings of the maple—but many plants have evolved to make use of animals, especially birds and mammals, to disperse their seeds. From an animal's viewpoint seeds, with their protein content, are nutritious food, but such consumption is a loss to the plant. Evolution has therefore led to the toughening of the seedcoat so that it may pass through the animal's gut unharmed and be deposited—with a little quantity of fertilizer—wherever the animal then happens to be. There is of course nothing in this for the animal and there would be selection against eating inedible seeds. Various solutions to this problem have evolved. One consists of the production of vast numbers of edible seeds so that the occasional seed escapes digestion; something of a lottery with long odds. Another evolutionary path is to produce a reward other than the seed for the animal: to develop a fleshy fruit containing one or more seeds with a hard covering, for example peach or tomato. The fruit is digested and the seed is either rejected or passes through the animal unharmed. In some plants the effect of the passage through the gut of an animal is necessary for subsequent germination—and it may be only a particular animal. Fruits for eating will generally be advertised by bright colours and, more rarely, by smell. Some tropical plants have evolved a specialized relationship with certain birds and their fruits are not particularly conspicuous.

Nuts are another strategy for seed dispersal. The hard, outer case ensures that only a limited number of animals will be able to open them—they will often take them away to open them and then drop them *en route*. But some nuts will be deliberately moved and spared. This happens because each nut contains so much food and as there are often so many of them at fruiting time, the animal does not need to consume them all at once; it is worthwhile storing them for later consumption. Storage generally involves some form of burial and as the animal always fails to find all the nuts some are well placed for germination. For instance jays and nutcrackers bury acorns and squirrels bury acorns and a variety of other nuts. In the Amazonian rainforest a rodent, the agouti, eats and buries tropical tree fruits to which it is attracted by the sound of them falling to the ground; often it opens the outer shell to get at the seeds or fruits inside, dropping some of these. If these fruits are not found and opened, the seeds within are doomed— brazil nuts, a number of which are held in a cannonball-like fruit, are a spectacular example. Thus fruits will evolve to be large, making plenty of noise as they fall, thus ensuring that they will be found. It may be that some of these large fruits evolved dispersal mechanisms with large, now extinct, animals as the agent. The more fruits that are available at the same time, the more likely many fruits or the individual seeds will be stored and lost, which is clearly to the tree's advantage. Thus nut-producing trees often produce large crops in some seasons, 'flooding the market', and only a few nuts in others. Furthermore any specialist fruit pest (likely to be an insect) will have its population squeezed in the years when the crop is light. Remarkably, all the trees in the same area will be synchronized; this is termed 'masting' and is particularly marked in beech trees. In forests with masting trees, the changes in the numbers of many animals are often governed by the fruiting cycle of the trees, so tight is the link that has evolved.

Other plants evolved a rather more one-sided relationship with animals for distribution of their seeds. Mostly they developed hooks on the outside of the fruit, burrs which become attached to the fur or feathers of passing mammals or birds and thus can be carried long distances. A few plants such as the mistletoe berry have sticky fruits. Several birds like the mistle thrush in Europe and the mistletoe-bird in Australia are particularly partial to them. After squeezing the berry and eating the flesh, the bird will clean its beak on the bark of a tree and thus plant the seed in the required position—for mistletoe is a parasite and grows not in the ground, but on trees. Perhaps the most one-sided relationship of all is that which sometimes occurs with the *Pisonia* tree. This grows on coral islands, in almost pure sand; it has a very sticky fruit and if this becomes attached to a relatively small bird, like a noddy tern, the bird may be unable to free itself and dies, so providing the seedling with excellent compost to start its growth in the barren habitat.

But how and when did this coevolutionary trail start? When did the flowering plants (angiosperms) appear? They are identified by a number of anatomical and reproductive characteristics of which the most reliable is termed double-fertilization. In this there arises from the pollen not only the male sex cell that pairs with the ovule of the flower, but another cell that fuses with two nuclei in the female flower giving rise to the seed's food store—the endosperm. This character, and to some extent the others, are difficult, if not impossible, to identify in the fossil record. However pollen is often well-preserved and the intricate structure (FIG. 9.13) permits identification—in more recent deposits down to the actual species of plant, so that a detailed knowledge can be gained of the flora at that time. Some pollen grains showing angiosperm characters that have been found in Europe, South America, and Africa from deposits dated 127–120 Mya (late Early Cretaceous) are considered to be the earliest definite evidence of flowering plants.

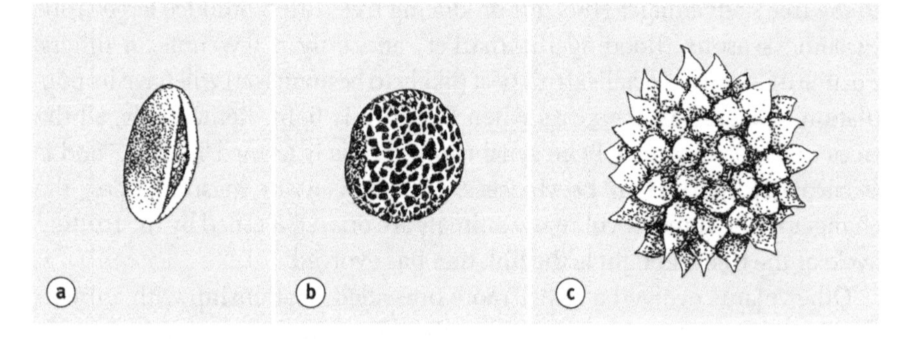

FIG. 9.13 Pollen grains of flowering plants: (a) horse chestnut; (b) lily; (c) ragwort

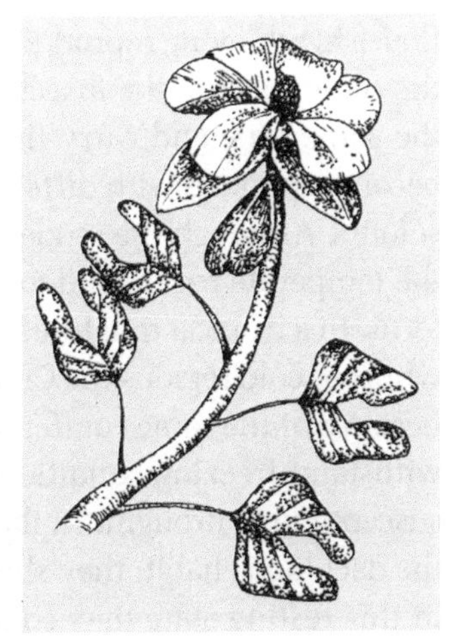

FIG. 9.14 An early flowering plant, *Archaeanthus* (after Dilcher and Crane, 1955)

It is believed that the earliest insect-pollinated flowers would have been bowl-shaped, like a modern magnolia, buttercup, or rose; their open structure means that they could have been pollinated by the wind until adaptations led to insects being the more important agent. The earliest fossil of this type is *Archaeanthus* (FIG. 9.14) found in mid-Cretaceous (around 97 Mya) rocks in Kansas. It is considered to be closely related to the magnolias. Beetles visit such flowers to feed on pollen and are likely to have been early pollinators. It is interesting that whereas most butterflies and moths have their mouth-parts formed into a tube-like proboscis through which they suck up nectar, the most primitive moths (Micropterygidae)—which have remained virtually unchanged since the Cretaceous—have mandibles and may actually chew pollen. This clearly represents an early stage in insect pollination before moths evolved to 'spare' the pollen of the plant. Today these moths may be found in buttercup and other bowl-like flowers in most parts of the world. Some of the closest evolutionary relationships are between bees and plants—these also have a long history as a bee has been found embedded in Cretaceous amber.

We can envisage that special relationships between flowers and their pollinators were already evolving by the end of the Cretaceous. But many early plants were wind pollinated, such as those related to the plane (or buttonball) tree. Fossil evidence suggests that these early angiosperms occurred in unstable soils along river margins and similar locations; they were 'weed trees'. A possible relationship between such a flora and dinosaurs has been suggested above. It is also interesting to note that the plane tree, like the maiden-hair tree, another survivor today from close relatives of earlier epochs, is noted for its resistance to sulphur dioxide and other pollution—an advantage in periods of volcanic activity.

By the end of the Cretaceous the flowering plants (angiosperms) were
ess abundant, than the conifers and their relations (gym-
the latter, which rely on wind dispersal, flowering plants
buting their pollen and seeds. Thus they could grow and

flourish, achieving reproductive success, some distance from other plants of the same species; the insect, bird or bat would search out another plant of the same type and carry the pollen. The canvas of the vegetation would become speckled with different greens and points of white and even other colours. Although the modern-day angiosperms have spread right through the temperate region and to the polar areas, one can still see these two patterns. In a tropical rain forest virtually every tree differs from its neighbour, whilst in cold regions the Cretaceous picture still holds—there will often be hectares of the same conifer. The leaves of many angiosperms are unable to withstand freezing conditions; they lose water quickly and so are equally susceptible to drought. In the Cretaceous period angiosperm trees evolved the deciduous habit: they shed their leaves in unfavourable conditions and in this resting state they could withstand periods of dryness, cold, or prolonged darkness. This ability to shut down could have been important in the face of conditions that arose at the end of this period. Only a few gymnosperms are deciduous.

Members of over 20 groups of plants that are living today have been identified in the fossil record of the Late Cretaceous. The modern members of these groups include water-lilies, heathers, mallows, myrtles, walnuts, birches, spurges and nettles. Unlike many other organisms that were present at that time these groups clearly survived the end of the Cretaceous.

The Cretaceous/Tertiary (K/T) extinction

It has long been recognized that the dinosaurs disappeared from the fossil record at the end of the Cretaceous period and, as more knowledge has been gained other organisms were seen to have been lost. Some like the plesiosaurs, pterosaurs, and rudists were becoming scarce before the end of the period, and this was even more true of the ichthyosaurs. Others like the mosasaurs, the freshwater sharks, and the ammonoids seemed to flourish right up to the end of the period and then disappear. Detailed studies reveal that many bivalve shellfish, lampshells, and sea-mats also became extinct and the microscopic plankton with calcareous shells suffered massively, with around 85 per cent of the coccolithophorid algae and all but a handful of the foraminifera having been lost. Chalk was no longer being laid down in the sea. The foundation of the major marine food chain that led, from these minute plankton through the ammonoids and other invertebrates up to mosasaurs, had collapsed. Some groups seem to have escaped: amongst the microscopic plankton the diatoms, with their siliceous shells and tough resting stages, survived as did many fish and deep sea and scavenging species such as the nautiloids.

On land it was not only the large animals that became extinct—the pterosaurs and dinosaurs were not alone in being affected. The mammals, most of which were small, lost some 35 per cent of their species worldwide. Plants were also affected. For example, in North America 79 per cent did not survive, and it has been noted that the survivors were often deciduous—they could lose their leaves and 'shut down'—while others could survive as seeds. As in the sea it seems that on land one key food chain collapsed: the one with leaves as its basic raw material. These leaves were the food of some of the mammals and of the herbivorous dinosaurs, which in turn were fed on by the carnivorous dinosaurs. Furthermore it is most likely that these large dinosaurs had slow rates of reproduction, which always increases the risk of extinction. Crocodiles, tortoises, birds, and insects seem to have been little affected. The two first named are known to be able to survive for long periods without food and both can be scavengers. Indeed, with the deaths of so many other animals and with much dead plant material the food chain based on detritus would have been well supplied. Many insects feed on dead material; furthermore most have at least one tough resting stage. In unfavourable conditions some may take a long time to develop: there is a record of a beetle larva living in dead wood for over 40 years before becoming an adult. Some birds were scavengers, but the survival of many lines, as suggested by studies based on the molecular clock, is a puzzle.

What happened in the biological story just after these extinctions—what is found in and just above the boundary layer between the deposits of the Cretaceous and those of the Tertiary, termed the K/T boundary? For a very short period the dominant micro-organisms in marine deposits were usually diatoms and dinoflagellates. The latter are protists with two hair-like flagella, one in a groove around the centre of the cell. But the important feature for the survival of both these groups was the ability to form protective cysts that rested on the sea floor. Above these, in the later deposits, are the remains of other minute plankton, but the types are quite different to those of the Late Cretaceous (FIG. 9.15). In terrestrial deposits a sudden and dramatic increase in fern spores marks the boundary in many parts of the world; ferns, as was noted from their abundance after the Permian extinction, are early colonisers of barren landscapes. The fern spike, as it is termed, has been found also in some marine deposits (such was the abundance of fern spores blown around the world), and it occurs in exactly the same layer of deposit where the plankton disappear. We can conclude that the major marine and terrestrial events occurred simultaneously.

Many theories have been put forward for the extinction of the dinosaurs, but most of them can be dismissed. Since 1980 there have been more focused, but still controversy-ridden, investigations. In that year Luis and Walter Alvarez and colleagues from the University of California published their research on the amounts of various metals in the boundary between Cretaceous and Tertiary rocks (K/T boundary) in Italy, Denmark, and New Zealand. They had found, accidentally (like many great discoveries in science), that a rare metal, iridium, suddenly became very abundant exactly at the boundary and then slowly fell away. This phenomenon, known as the iridium spike, has now been identified in K/T boundary deposits in over a hundred other sites in the world (FIG. 9.15). Iridium occurs in meteorites and in volcanic material, but in the latter case it is accompanied by elevated levels of nickel and chromium. These other metals are not especially abundant at the K/T boundary. The Alvarezes (father and son) concluded that the iridium spike was due to a large asteroid (now often referred to as a bolide) that struck the earth 65 million years ago. They calculated, from the amount of iridium spread around the world, that the asteroid would have been some ten kilometers in diameter and would have made a crater 150 kilometres wide. The effect of this would have been not unlike that of a nuclear war—it would have led to a 'nuclear' winter. In the immediate vicinity the underlying rock would be melted and debris of all sizes flung into the air; an enormous cloud of dust would envelop the earth cutting out sunlight and, if the impact had been in the sea, huge tidal waves (tsunamis) would sweep onto near and distant coastlines.

This catastrophic scenario was by no means generally accepted. Many other scientists, particularly those who worked with fossils (palaeontologists) argued—and some still do—that the extinction had not been sudden. They suggested that different organisms faded out at different times; changes in the geography, due to continental drift, led to changes in the climate and sea levels. This was the gradualist argument; other suggestions were more catastrophic, but implicated events of terrestrial origin. The most plausible proposal of these was that there had been a prolonged and extensive period of volcanic activity (as is now believed to have been important for the Permian extinction): the Deccan Traps, great outpourings of volcanic rock, in India and under part of the Indian Ocean, were dated from

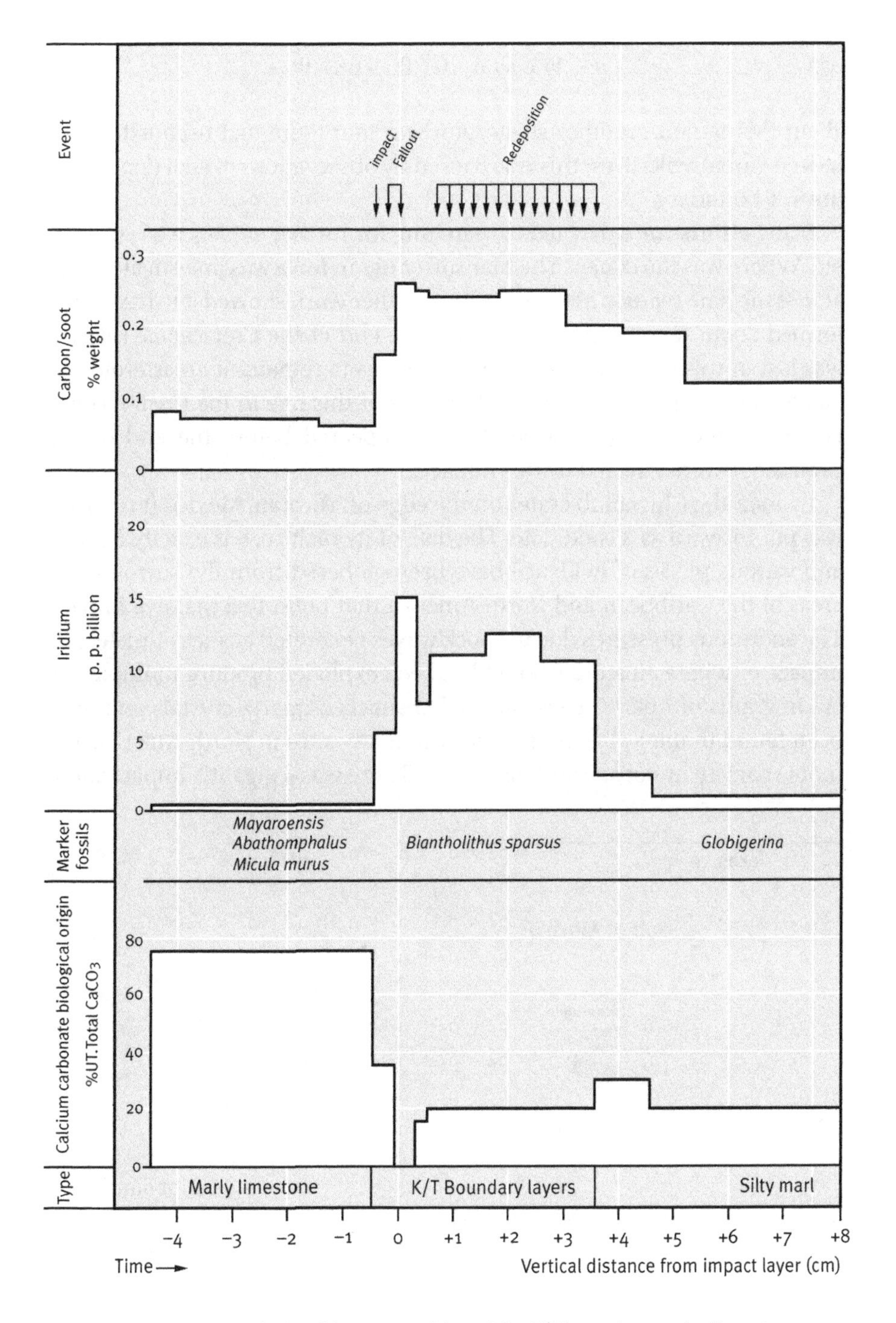

FIG. 9.15 An analysis of the composition of the K/T boundary and adjacent rock layers at Elengraben, near Saltzberg (modified from Preisinger *et al.*, 1986)

about this period. Another suggestion was that there had been extensive, indeed global, wild-fires; this was backed by observations of soot deposits at the K/T boundary.

Many efforts were devoted to searching for further evidence of an asteroid. Where was the crater? The Manson crater in Iowa was investigated, and at first the date seemed about right, but further work showed it to have been formed about ten million years before the end of the Cretaceous. Interest was lost in this crater, but the question seems to remain: if an asteroid has had a major impact, where is the footprint of this one in the fossil record? Perhaps some of those species that disappeared before the end of the Cretaceous were affected by this impact.

In 1992 the Chicxulub crater on the edge of Yucatan, Mexico (FIG. 9.16) was put forward as a candidate. The date of its melt rock is exactly 65 Mya and various pieces of evidence have been gathered from the surrounding areas of the Caribbean and North America that point to a massive impact. The enormous pressures due to shock waves generated at sites of meteorite impact, or where an atomic bomb has been exploded produce multiple layers in grains of quartz. These are called shocked quartz crystals and have been found in the K/T boundary layer in many sites in North America and more sparsely in Europe and the Pacific. There was a gigantic impact and it

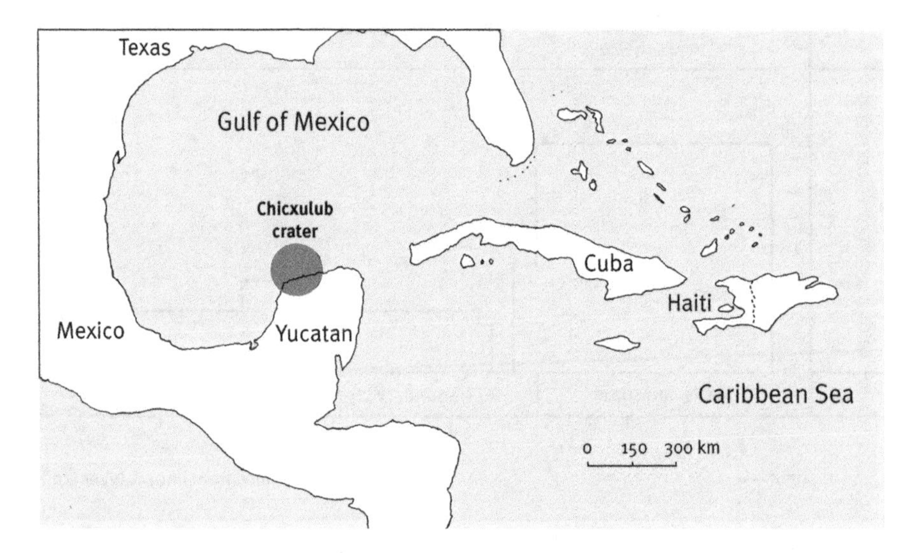

FIG. 9.16 The position of the Chicxulub crater (circle) in relation to present geography

was clearly in or near North America. When molten rock is splashed out into the air small spherules of glass and larger bead-like objects (called tektites) are formed; these have been found as far away as several hundred kilometers to the east in Haiti and to the north-west in Colorado and New Mexico. Ejecta, rock thrown out of the crater, which has the particular features of the underlying rock in Yucatan, has been found as far afield as Central Canada. Evidence of a tsunami has been found in Texas and north-eastern Mexico, this giant wave having swept over the land and then drained back to the Gulf of Mexico so that leaves are found in marine deposits. Studies on the crater indicate that the asteroid had a diameter of ten kilometers, which was the diameter calculated by the Alvarezes from the global quantity of iridium. Material would have been blasted out to a depth of 14 kilometers; the original crater was at least a 150 kilometres across, perhaps considerably more.

After this work, and other studies on rare elements and on rock form and composition, there can now be little doubt that an asteroid hit the earth in the region of Yucatan at the very end of the Cretaceous. The thermal blast would have destroyed life over a large area, but the effects would have been global. The cloud of dust would have cut out the sunlight for at least several weeks and reduced its strength for several years. This 'lights out' scenario would have closed down photosynthesis. We have noted already that it was the two food chains based on photosynthesis that were most affected; one in the sea, that originated with plankton, and one on the land, that commenced with leaves. The survivors in the sea were cyst-forming diatoms and dinoflagellates and, on land, plants that were deciduous or had resistant spores or seeds—they could shut down. The reduction of sunlight may have been intensified by the soot from wild-fires: the vegetation that had died would become dry and susceptible to ignition by lightning. Much of the rock that was vaporized was limestone and this would have produced a quantity of carbon dioxide which dissolved in the water vapour and then would have fallen as acid rain. We can postulate that this would have been another blow to those microscopic plankton with shells made of calcium carbonate; however freshwater life might be expected to be more damaged by acid rain and, though freshwater sharks were lost, most groups seem to have survived. So perhaps the rain was not so acid, or most freshwater bodies were well buffered.

The conclusion that must be drawn is that the Chicxulub asteroid would have had a dramatic effect on life and that these effects would have occurred over a short time. One research worker has claimed that a study of plant remains from the time shows that the impact occurred in June. But questions remain: was this asteroid enough to cause the global extinction that occurred at this time and how does it explain that, in places in the fossil record, it seems that different organisms became extinct before the impact, at times that can be separated on a geological scale. It is indeed one of the problems in this debate that the time-scales are so different. A depth of a centimetre in a deposit may represent a thousand years and this layer may have been stirred up by burrowing animals and storms. Geologists can be proud of estimates whose accuracy is given to fractions of a million years. Survival of organisms has to be considered in terms of generation times and, in these circumstances, for how long an organism can 'shut down'—either in a specific resting stage (as a cyst or seed) or simply in what might be termed suspended animation. From the biological viewpoint the fatal effects of an asteroid would extend for no more than several years—the time it would take for at least some sunlight to break through. This would save those whose resting stages were adequately long and could live on in the now greatly changed environment. One notes again how many of the survivors were carrion and detritus feeders or part of that scavenging food chain, such as insectivorous mammals themselves living on insects that themselves lived on fungi or on material being broken down by microorganisms. Certainly initially, in the areas away from the immediate fireball, they would have had ample food supplies.

There are some who believe the Chicxulub asteroid is not the explanation for many of the K/T extinctions, though I judge that such an asteroid must have had wide-reaching effects. However there is also good evidence that some extinctions did occur before 65 Mya and this is where the other explanations—climate and sea-level change and volcanism—are relevant. A combination of such factors brought about the spectacular extinction at the end of the Permian and undoubtedly they would have stressed the ecological processes at the end of the Cretaceous. The volcanism associated with the formation of the Deccan Traps would, one would think, be significant, and there have been various estimates of the age of these great areas of basaltic rock. In 1999 geologist Claude Allegre and colleagues of the Institut de Physique de Globe, Paris, made an estimate of 65.6 Mya, with an accuracy of plus or minus 300,000 years. They based this on measurements made over a wide area and at different depths and it indicates that in biological terms this volcanic event occurred well before the asteroid impact, which was 300,000–900,000 years later: its effects would have preceded those of the asteroid.

Though it seems that this extinction may have been unusually dominated by one factor, we can conclude that on present evidence it was, like the earlier mass extinctions, a concurrence of changes that challenged life in new ways. This challenge was on a grand scale, a scale that has not occurred since. The dominant forms of terrestrial life that developed in the wake of this extinction—mammals, birds, insects and flowering plants—are those with which we live today.

CHAPTER 10

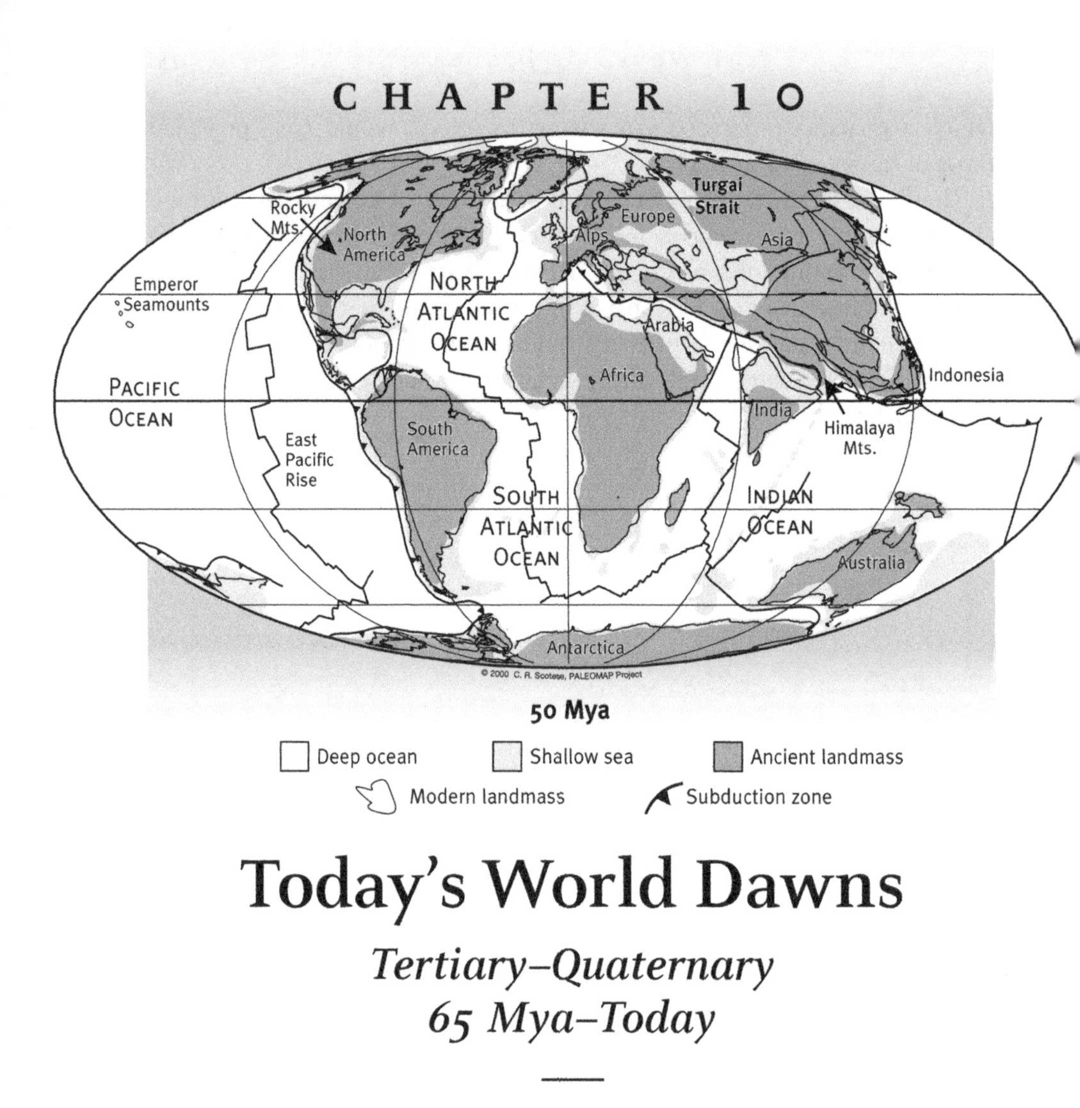

Today's World Dawns

Tertiary–Quaternary
65 Mya–Today

THIS period which brings us right up to the present, is often referred to nowadays as the Cenozoic. The first part of it, the Tertiary, is by far the longest component, generally being taken as extending until 1.81 million years ago. It has itself been divided into a number of periods: Palaeocene (to 55 Mya), Eocene (to 33.7 Mya), Oligocene (to 23.5 Mya), Miocene (to 5.2 Mya), and Pliocene (to 1.64 Mya). The majority of the second part, the Quaternary (1.64 Mya–10,000 years ago), is generally known as the Pleistocene.

At the opening of the Tertiary period, the Tethys Ocean still lay between Laurasia and Gondwana, but it was being squeezed by the movement northwards of India and of Arabia and Africa. As the plates pushed against each other the land was forced up, leading to the birth of the Alps and Himalayas. Eventually the continental land masses came into contact virtually squeezing the Tethys Ocean out of existence. After the sharp fall in sea level at the end of the Cretaceous (see FIG. 4.1), the level rose again in the Palaeocene and large parts of southern Europe and south-west Asia were covered by a sea, called the Paratethys. Its continuation north through central Russia, east of the Urals, is known as the Obik Sea and this linked through the Turgai Strait to the Arctic Ocean. Later the Paratethys and Obik seas gradually shrank and all that remains today are Lake Balaton and the Black, Caspian, and Aral Seas, with the Mediterranean opening up to the south. Its link through to the Persian Gulf (as it is now) was cut about 19 Mya. Thereafter African animals could cross into Laurasia and vice versa. When the sea level was at its highest it is believed a trans-Saharan sea spread down from the Mediterranean, possibly as far as the Gulf of Guinea.

The continents of Gondwana had largely separated by the start of the Tertiary, though Australia and Antarctica were close until the Oligocene, about 30 Mya. Laurasia divided earlier, in the Early Eocene when, as the North Atlantic widened, North America and Greenland split from Europe (this widening is still occurring at a rate of about one centimetre a year). The Arctic Ocean was now linked to the Atlantic and no doubt there was a major change in the oceanic currents that have such a profound effect on the climates of adjacent continents.

Then in the Miocene, in a period when the sea level was low, the Straits of Gibraltar were closed and, with evaporation, the Mediterranean became a very salty sea until, around the start of the Pliocene—about five million years ago—as global sea levels rose again, the Atlantic broke through the barrier and refilled the Mediterranean. Thus life in this sea today is a reflection of that of the Atlantic, rather than the richer fauna it probably inherited from the Tethys and Indian Oceans.

Although North and South America had been linked in the Cretaceous and Paleocene by a somewhat disrupted land bridge, the Greater Antilles Island Arc, this link was broken when the Arc moved north-east on its tectonic plate. It was not until about 30 million years later, in the mid-Pliocene (some 3.5 million years ago), that the present link was established. During the period of isolation the mammals on the two continents had evolved so that the faunas were composed of very different animals but, following the mixing (the so-called Great American Interchange), there was an overall reduction in diversity. The South American fauna suffered the greater extinction and fewer southern animals moved north (12 genera), than northern ones moved south (27 genera). Furthermore of the 12 which moved North only three exist today, whereas the invaders from the north were much more successful, and as they established themselves and diversified, new genera evolved. This picture has been interpreted as showing that the South American fauna, which contained many animals that could be described as primitive, was out-competed by the more efficient northerners. However, it has been pointed out that the South American mammal fauna was showing some signs of diminution in the Early Pliocene; so it is possible that the environment was changing and the northern animals merely occupied vacant niches. These two views are a reflection of the debate, often passionate, on the role of competition in community structure. In my view both factors are likely to have operated, although we should not think of competition as involving two animals fighting it out (except in the case of a carnivore and its prey); rather that the successful feeding of one species, would, for example, reduce—in some way—the food of another.

Temperature changes over the Cenozoic have been substantial (FIG. 10.1). Levels declined from the middle of the Cretaceous until early

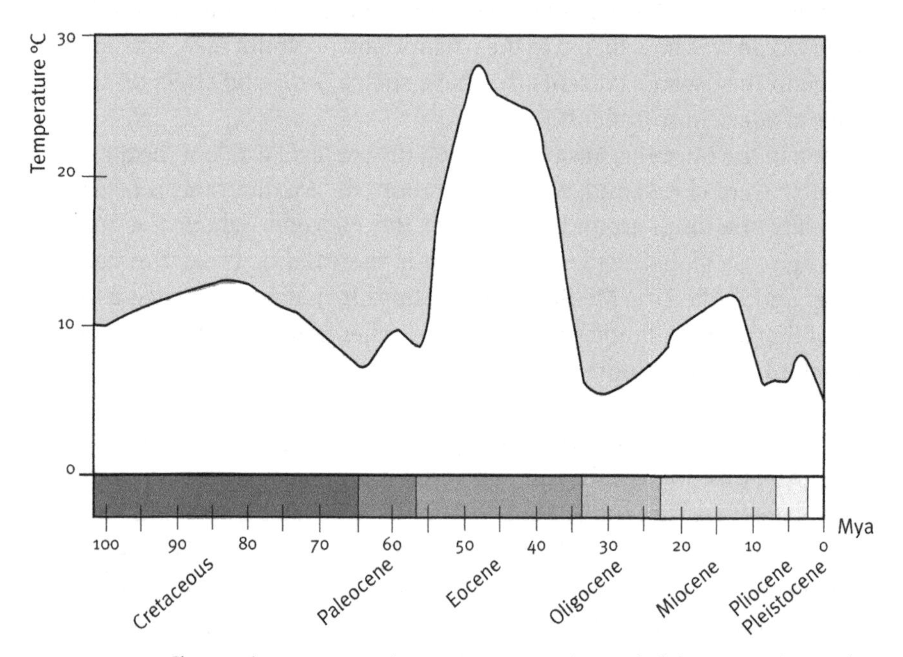

FIG. 10.1 Changes in mean annual temperature over Cenozoic (after Novacek, 2000)

in the Palaeocene, after which there was a more or less sustained rise in global temperature for about 17 million years until the mid-Eocene, some 48 million years ago. At this time the poles were free of ice and subtropical vegetation extended to 60° north, the present latitude of Oslo. Studies of fossils in the London clay have shown that the early-Eocene flora of southern Britain was similar to that found in south-east Asia today. This was the last really warm period, as there was a major cooling from this peak and by 38 Mya, towards the end of the Eocene, ice was developing in the south polar region. The Oligocene was cold, but temperatures rose again somewhat during the mid-Miocene, around 14 Mya, falling again to give a yet smaller peak in the late Pliocene, 3 Mya. These warm periods were associated with the melting of a large part of the southern polar ice with the consequential rise in sea levels and the advent of a more humid climate. At the start of the Pleistocene, temperatures were low and ice sheets were also forming around the North Pole.

Immediately after the K/T extinction the flora in much of the world was dominated by ferns. Studies of leaf fossils in North America by Jack Wolfe and Garland Upchurch of the US Geological Survey have shown that the ferns were followed by species that are typical early colonists of disturbed sites. These are known as 'early successional' because they are the first in a series of plant communities that will succeed each other provided the habitat is left unchanged; the final community is termed the 'climax'. About 1.5 million years after the K/T extinction the form of the leaves changed, with serrated edges and lobed leaves disappearing (FIG. 10.2). The new leaves had smooth continuous margins and ended in a point—the drip tip. These leaves are characteristic of trees growing in warm, wet climates, and this is what would be expected in the absence of polar ice, and the rising sea level. The drip tip helps the water to drain away from the surface of the leaf, for these are often evergreen and, living for a long time, they need to discourage the growth of fungi, algae, and lichens on their surface by keeping it as dry as possible. Some of the plants that we keep in our homes and offices have well-marked drip tips (e.g.rubber tree), which shows us that their natural home is in the tropical rainforest.

Nature at its richest

During the Tertiary period the main bands of vegetation, which we still have today, became clearly established: the tundra, coniferous forest, deciduous

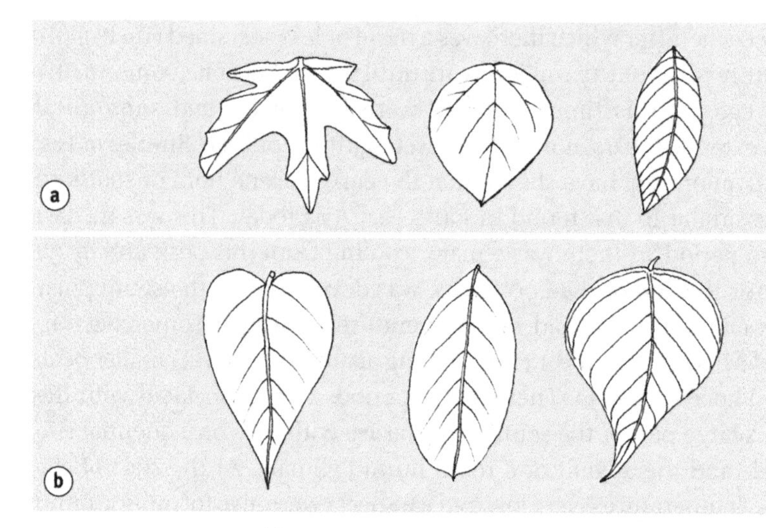

FIG. 10.2 Leaves: (a) from early Tertiary; (b) rain forest trees showing drip tips ((a), after Wolfe and Upchurch, 1987)

forest, grassland, and tropical rainforest. By far the richest, in terms of its diversity of plants and animals, is the last of these. A tropical rainforest is defined as having a closed canopy—that is, the trees meet together—and it is multistoried, in that the trees spread their branches at several levels (FIG. 10.3). Those trees that stand out above the rest may be well over 50 metres in height, but below them the various layers are sometimes quite distinct. To study life in the rainforest towers have been erected and as one makes one's way up—or down—one is often struck not only by the different visual features, but by the distinct smells of each layer.

The diversity of tree species in rainforests varies, and this can be so great—as it is in Peru—that 300 different species may be found in a single hectare. But trees are by no means the only component of rainforest vegetation. There are climbers, or lianas, that can grow up to the highest canopy, depending on the trees for their support. Then there are epiphytes, plants that grow not in the soil, but on the trunks and branches of the trees: mosses, orchids, ferns, bromeliads, and others. The number of species of plant in these two categories may equal or exceed the number of different trees.

Animal diversity greatly exceeds that of plants. It has been estimated that the number of species of beetle associated with just one species of tree is between two and four hundred and about one tenth of these live only on that type of tree. Quite the most abundant insects in a rainforest are ants, of which there are often some tens of species, many with different habits—some living on the trees in nests of leaves woven together, others nesting in the ground, some in rotten wood and yet more kinds with other habits. Some of these will fight each other to hold particular areas: the forest has been described as an ant mosaic. Almost all the other animals are influenced by the ants: some will be harboured and protected, more will be attacked and often eaten. In South America there are leaf-cutting ants that may destroy more foliage than all the other animals put together; they take the portions of leaf back to their nests where they grow on it a special fungus for food. (This is a rather striking example of the need for animals to depend on micro-organisms to obtain their nourishment from leaves, p. 102). Ants have many other relationships with plants which often bring mutual benefits. Some seeds have a structure on the outside that is very attractive to ants; they will pick them up to carry back to the nest and so benefit the seeds by distributing them. Other ants live in epiphytes, in special chambers, and so gain shelter while the debris they produce in the chamber provides the plant, which does not have its roots in the soil, with nitrogen.

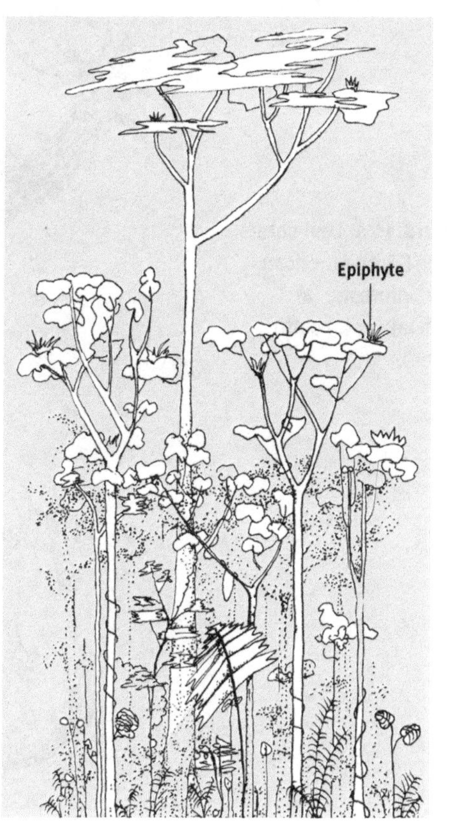

FIG. 10.3 Diagrammatic representation of the layers of the canopy, with epiphytes and climbers, in a rainforest (modified from Richards after Oldemann, 1974)

The diversity and brilliance, if you have the rare opportunity of seeing them, of rainforest birds is outstanding. During a walk in a forest in New Guinea with Jared Diamond—an expert on the birds there (and many other things)—he identified, in the space of four hours, over 120 different bird songs, although we saw very few of the songsters. Many rainforest birds are insect feeders; some, like the parrots, are fruit eaters; while others feed on mammals and other birds.

In Central and South America the Harpy Eagle, with a wing span of two metres, can capture large monkeys and sloths. Tapirs, certain wild pigs, deer, large cats (like the jaguar), large rodents and, in parts of Central Africa, elephants and okapi, live on or near the forest floor, but the most abundant of the larger mammals—the monkeys—live in trees, often only coming down to the ground to drink or lick rocks for particular minerals. The latter seems especially important to those species that feed mainly on leaves; the sole food also of the sloth and of a strange bird, the hoatzin (FIG. 10.4(a), see also FIG. 7.1)—whose young has claws on the front of the wing. Fruits are the

FIG. 10.4 Leaf eaters in South American rainforests: (a) hoatzin; (b) sloth

major food of most monkeys and they travel in groups over the forest, searching out those trees in fruit; this demands co-ordination between the members of the group, whilst a memory from season to season and knowledge passed on from generation to generation is a great advantage. Little seen, but enormously abundant, are the smaller animals, especially rodents; hollows in trees will often reveal a mini-zoo of inhabitants.

All these organisms are linked through complex food webs and as these operate over millions of generations special adaptations evolve, to eat better or to improve the chance of not being eaten; that is, to increase the likelihood of survival and reproduction, the measure of evolutionary success. Although there is such a richness of species in rainforests, many are specialists so that a pair of organisms will evolve together, and this reciprocal interaction is termed co-evolution. Many co-evolved adaptations are remarkable and the examples in rainforest organisms are often particularly striking.

An example of a rich co-evolutionary web that has been well studied by Larry Gilbert of the University of Texas and colleagues, is that associated with *Heliconius* butterflies in Central America (FIG. 10.5, overleaf). It can be only briefly outlined here, but even these few details give an idea of the number and complexity of the links between the thousands of species in a rainforest. Tragically many of these species are being driven to extinction by man's destruction of their habitat, often before we really know of their existence.

The food plant of the *Heliconius* caterpillar is the passion flower vine. This is sparsely distributed in the forest and only occasionally produces the young shoots that the caterpillars need. The butterflies need to live for a long time (perhaps as much as six months) to find enough sites in which to lay all their eggs. For this they need protein food as well as nectar and, unlike most butterflies, they collect a ball of pollen on their 'tongue' into which they dribble and then suck back the soup of partly digested pollen. They obtain the pollen and nectar they need from certain wild cucumber vines. The male and female flowers are separate; many male flowers are borne on a single shoot and though each one comes out for only a day, the whole cluster of flowers offers an abundant source of pollen for several weeks. By being in one place this ensures that the butterfly visits regularly and so carries the pollen on their rarer stops at female flowers where they only get nectar. The female butterfly therefore has to find scarce passion vines growing new shoots and equally scattered cucumber vines. They appear to fly round the forest in something of a route, day after day, checking the passion vines and feeding at the cucumbers, with a gliding, low energy-cost flight.

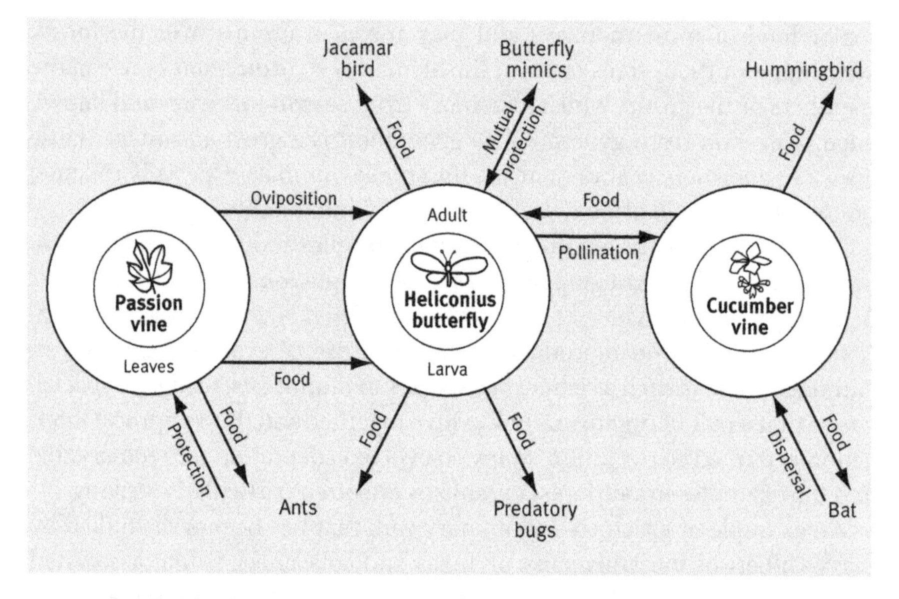

FIG. 10.5 The dominant links in the ecological web around the *Heliconius* butterfly in central American rain forest. The arrows point to the organism gaining the benefit from the relationship. The web extends, virtually infinitely, through all the links of the other organisms

The growing shoots of the passion vine are not very large, generally sufficient only to sustain one caterpillar. If two caterpillars meet, the larger will consume the smaller; therefore natural selection has led to female butterflies avoiding laying on a shoot that already has an egg on it. Passion vines have evolved outgrowths of the shoot that look like butterfly eggs, thereby discouraging egg-laying. If an egg is laid and the caterpillar hatches, it will usually destroy the shoot; but the caterpillar has enemies, of which the most abundant are ants. The plant has evolved to attract ants by having extrafloral nectaries that secrete a sugary juice on the base of the leaves. If while coming to feed on this syrup an ant comes across a caterpillar the latter will be attacked and carried back to the nest for the ant larvae to eat.

Heliconius butterflies are the herbivore that has the major impact on passion vines in their natural habitat—hence the vine's targeted evolutionary response. This lack of herbivores is due to the presence in the vine of secondary plant substances, that are distasteful or even poisonous. The butterfly has overcome this defence and actually stores these toxins in its body so that it is unpleasant to eat and its bright red, orange, and black colours are to warn birds and mammals to avoid it. However, this has to be learnt by hard experience, therefore the more insects sporting these warning colour patterns the greater the chance any individual has to avoid being the sacrificial victim. This evolutionary advantage in being part of a crowd has led to the evolution of mimicry; there are other species of butterfly that mimic the *Heliconius* so closely that they are hard to tell apart. Most of these butterflies are also distasteful in a variety of ways, so mimicry is a mutually beneficial adaptation and the group of insects resembling each other are referred to as being in a Müllerian Ring—named after the naturalist who first described it. However—evolution ensures there is always a 'however'—there is one group of birds, the brightly coloured Jacamars, that do not find all the members of the *Heliconius* ring distasteful. These birds have particularly long beaks, ideal for manipulating a butterfly so as to take a close look and see if it is edible.

It used to be considered that this great variety of life in rainforests was a reflection of their stability, that they remained unchanged over countless years. This is now known to be incorrect. On the shortest scale, both of time and of size, these forests are always changing because the gigantic trees are both shallow-rooted and relatively short-lived, the heart often rotting; they fall and when they do so, the gap is recolonized by a succession of plants. These gaps open at random across the forest and what we see is actually a mosaic of vegetation all in different stages of succession; this plays a major role in the maintenance of diversity. Contrary to earlier belief rainforests will burn; large areas of the Borneo forest were destroyed by fire a few years ago. Today such fires are usually started by man, but lightning could be the cause in particular circumstances. Robert Johns, formerly of the University of Lae, considers that, as a result of fires, rainforests in New Guinea are mostly less than 100 years old.

As already mentioned, fossil leaves show that plants with rainforest characteristics occurred from fairly early in the Palaeocene; such vegetation spread as the climate became warmer and wetter until the mid-Eocene, after which there was a waning and waxing in the extent of rainforests. When the climate in the tropical regions became drier and cooler after the Middle Miocene and again in the mid-Pliocene, their area was probably at its most restricted. The forests would have retreated again during the various ice ages that occurred in the Pleistocene. Instead of great unbroken areas of forest it would have been limited to isolated patches termed 'refugia'. It is believed that these periods of fragmentation were the cause of the rich diversity. During the period when the forest was patchy, populations would be isolated in the many refugia; these would follow different evolutionary paths so that, when in warmer and wetter times, the forest spread and the animals and plants were brought in contact again, they would not interbreed; they would have become different species. The whole series of ice ages would cause this 'species pump' mechanism to operate many times, behaving rather in the way of compound interest. This theory was developed in particular in relation to South America and it was proposed that upland areas, retaining damper conditions, were the refugia, the intervening lowlands being arid and inhospitable to rain forest denizens. However, as a result of examining the pollen in lake deposits in Amazonia, Paul Colinvaux of Ohio State University, has suggested that it was cold weather rather than dryness that was the factor limiting refugia and so they would be found in lowland regions rather than the mountain areas which would have been too cold.

Grasslands get going

The dry periods encouraged the development of a new type of vegetation, grassland: some without trees (prairie or steppe), and some with scattered trees and bushes (savannah). Grass pollen is first found in some quantity in the Oligocene and it may be that some grasslands were then established in South America; but in most of the world it seems that it was not until the mid-Miocene, with its warm climate, that typical grasslands appeared. Initially these contained a large component of shrubs and trees and it seems that, at least in North America, pure prairie grassland was not widespread until the early Pleistocene. Associated with the expansion of grasslands has been the evolution of many types of animal: browsers (eating shrubs and trees) and grazers (eating grass), together with associated predators and scavengers. Grasslands can sustain very large populations of these animals and this is in part due to the unusual growth form of grasses. The new shoots that grow when an older one is eaten (or cut) arise from buds which are protected, being at or below soil level. When one shoot is lost, more than one may grow in its place until there is a great density of shoots, as we see on a garden lawn.

Many groups of animal, probably including humans and our relatives, have had their evolutionary history affected by the emergence and spread of grasslands. Large herbivores whose history is intimately linked to that of grasslands are the ungulates, of which there are two groups: those with an even number of toes, two or four (the artiodactyls) and those with an odd number of toes, one or three (the perissodactyls). The former are the cloven-hoofed animals and include cows, sheep, goats, deer, giraffes, camels, and llamas; they are also known as ruminants, from the name, rumen, of one of the three or four pouches into which their stomach is divided (see FIG. 7.11). Initially the food is stored in the rumen where cellulose-splitting bacteria start to break it down, after which it is regurgitated and chewed (chewing the cud). It is then reswallowed and, provided it is small enough, passed through the rest of the digestive system. If it is not small enough it will be regurgitated again. The perissodactyls consist of the horses, zebras, tapirs, and rhinoceroses. These animals concentrate their digestion of herbage in the hind gut and so can process a high volume of food rapidly; in this way they are well adapted to surviving on food with a high content of fibre, like straw or the leafy twigs of bushes. Chewing the cud is a slower process, but more nutrients are absorbed. Ruminants can therefore survive on less food, but it has to be of better quality—they cannot cope with a large amount of fibre which tends to choke up the relatively small openings in the stomach—and they are also able to extract water more efficiently. For these reasons ruminants generally survive better than horses, and their relatives, in habitats where there is scattered but good quality food, such as patches of grass. Horses and rhinoceroses need to harvest a large quantity of food; it will pass through their gut more quickly and therefore they spend more time actually cropping the vegetation, for which their sharp front teeth are well adapted. They have an advantage over ruminants in habitats with ample, but fibrous, plant material, and adequate availability of free water, on which they are more dependent than ruminants.

When many of the land masses were still connected, animals that can be loosely termed ancient ungulates, existed in all parts of the world. By the end of the Oligocene they had disappeared from everywhere but South America. In Laurasia the odd- and even-toed ungulates had evolved from them. In South America these ancient ungulates persisted and evolved into many body forms, paralleling the evolution of the modern groups in Laurasia: some were superficially like camels, others like large bison, some more pig-like. A great many of these persisted until the mid-Pliocene (3 Mya) when their disappearance seemed to coincide with the establishment of the land bridge between North and South America and the subsequent invasion of other ungulates and new predators from the north. A few persisted until well into the Pleistocene and until the invasion of other predators from the north—humans (p.234).

The early odd-toed ungulates (perissodactyls) were small, about the size of a terrier dog, and gave rise to three main groups, all of which are living today although only one is at all numerous—the genus *Equus* (horses, asses, and zebras). This genus travelled from North America to Asia across the Bering Islands land bridge at the start of the Pleistocene and from there spread to Africa and Europe, evolving into the various horses, asses and zebras. Ironically horses became extinct in North America in the middle of the Pleistocene, so if it had not been for some individuals crossing from Alaska to Siberia, horses would only be known to us today as fossils and they could not have played their significant role in human history. In their heyday the three-toed horses were very diverse; over 30 species were found at one time or another in North America in the Early and Middle Miocene. These early horses are thought to have been mainly browsers and when the grasslands spread they decreased while the number of ruminants rose, reflecting perhaps the different efficiencies of the digestive systems of the two groups as mentioned above. At the same time the modern, one-toed horses evolved to be grazers.

The fossil record for horses is good and various trends can be followed. These include increasing in size, lengthening of the legs and feet, reduction of the lateral toes and emphasis on the middle toe (on which the modern horse walks), and widening of the front teeth (incisors), both top and bottom, to produce an excellent cutting machine. As animals showing more than one stage in this process apparently occurred at the same time, it seems that this was not a simple gradual transformation. There may have been a sudden spurt in one lineage, with others remaining little changed over the same period. This is, however, a controversial area and other interpretations are possible.

There were other groups of perissodactyls, two of which have members living today, namely the rhinoceroses and the tapirs (FIG. 10.6). Although some extinct rhinoceroses appear to have been lightly built and less than two metres in length, their evolutionary path seems to have tended towards a heavy build and large size. *Indricotherium*, often called *Baluchitherium*, found in the Oligocene and Miocene of Asia, was the largest. Formerly it was thought to have weighed up to 30 tons, making it by far the largest terrestrial mammal, but recent measurements on nearly a hundred specimens suggest a maximum between 15 and 20 tons; three times the weight of a large modern elephant. After a period of success various types of rhinoceros died out in the latter part of the Tertiary. One group, to which the five living

FIG. 10.6 The Malayan tapir, with young—a perissodactyl (odd-toed ungulate)

species belong, remained into the Pliocene, although at that time it became extinct in North America. The woolly rhinoceros was widespread in cooler regions in the Pleistocene and often figures in cave paintings from that period. It became extinct some 11,000 years ago and four of the five living species are currently in danger of following the same path.

The tapirs (FIG. 10.6) are in many respects the most primitive of living perissodactyls retaining four toes on the front foot and three on the hind. They were widespread until the last part of the Pleistocene, when they disappeared from northern lands surviving only in two areas—Malaysia and Indonesia, and Central and South America. Like the horses they are extinct in their evolutionary homelands.

The even-toed ungulates, the artiodactylids (FIG. 10.7), most of which can be classed as ruminants, are the dominant large herbivores today being by far the most diverse and abundant. Their evolution has parallels with that of the odd-toed ungulates. Both have shown a general trend of increasing size, which has several advantages. The larger animal has less surface area in proportion to the body, so that less heat is lost in cool conditions. A large animal is better able to defend itself from carnivores—a full-grown and healthy rhinoceros or hippopotamus is beyond the powers of any lion. A disadvantage is that it will normally have to cover a larger range to gather enough food and even so will probably have to broaden its diet; furthermore its increased weight puts more strain on the bones (p. 76) and to protect these it must be less agile. The evolution of running ability is a common trait in the smaller of these ungulates: the pronghorn can achieve 86 k.p.h. (55 m.p.h.). Such an

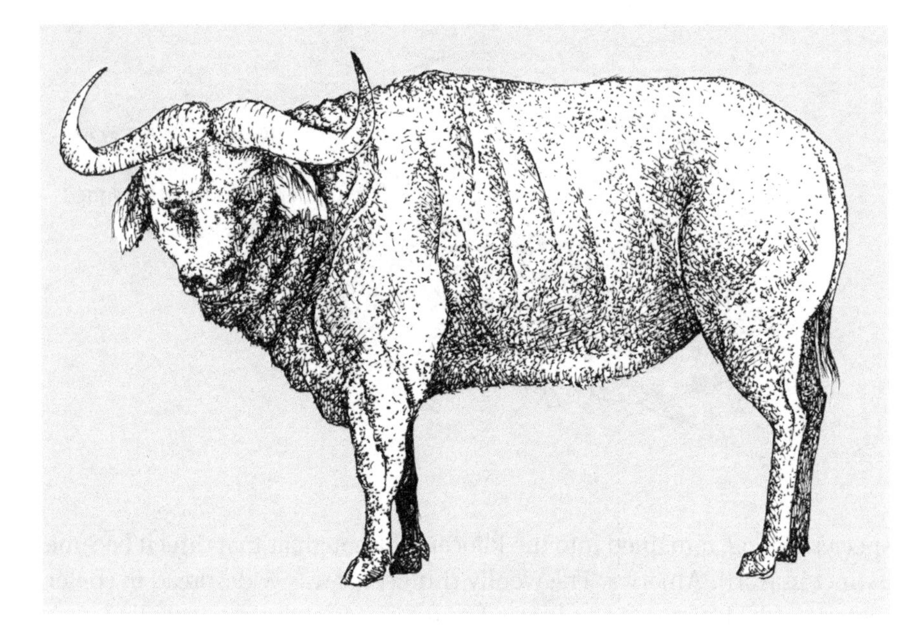

FIG. 10.7 The African buffalo—an artiodactyl (even-toed ungulate)

ability is another anti-predator device, important in a group whose evolution was tied to the development of a habitat without shelter—the grasslands—where one cannot hide from the enemy, and must rather flee. Also associated with this type of habitat is the tendency to form herds, a behaviour which has advantages in the battle against carnivores. Safety in numbers comes as several sensitive noses, pairs of eyes and ears are likely to spot an enemy before it comes too close. The members of a herd may be able to muster a collective defence: musk oxen will form an outward facing circle against wolves and buffalo may collectively charge a small pride of lions. A third trend has been the evolution of some form of horn, antler or tusk. These often play a major role in the social life of the herd, males in particular using them in combat, real or feigned, to achieve the leading position. Tusks are also used to dig in the ground for plant roots and other food. But whatever their other uses, all of them are of value as a last-resort defence weapon, a similar role to that suggested for the thumb of some ornithopod dinosaurs. This role was probably the initial driver in their evolution.

It is indeed interesting to compare the defence strategies of large mammalian herbivores with those of the dinosaurs. There are many apparent parallels, although with the larger dinosaurs being three or more times larger than the largest mammal and the average size of an adult being somewhat like that of an elephant, swiftness of foot was less easily achieved. Undoubtedly some of the ornithopods were the most nimble and they corresponded to the even- and odd-toed ungulates and lived in herds. *Triceratops* was the rhinoceros of the Cretaceous. The sauropod adult was like an adult elephant, and when healthy was beyond the range of any contemporary predator. The giant armadillos (glyptodonts), found at the end of the Cenozoic (see FIG. 12.2) resembled small ankylosaurs, some even having a mace-like ball on the end of the tail. There is no group analogous to the stegosaurs, but their peculiar structure was probably more related to thermoregulation than defence. These parallels emphasize the strong evolutionary force that predation exerts on herbivores.

The first fossils of even-toed ungulates (artiodactyls) are found in the Early Eocene and these evolved into a number of groups; by the end of that period and in the cool Oligocene, representatives of the pigs and peccaries appeared in Europe and North America respectively. Pigs (and peccaries) take a mixed diet of small animals, fungi, fruit and other plant material but they do not ruminate. Fundamentally they have changed little since the Oligocene, although there have been some large species, one of which in the Pliocene was as large as a modern rhinoceros. Some of the present species, like the warthog, feed almost entirely on grass, but in essence they are animals of damp forests, frequently visiting ponds and streams. A related group became more amphibious early in the Pliocene, and today these are represented by the hippopotami. Grazing at night, grass constitutes the major part of their diet though they also feed on aquatic vegetation. They have digestive pouches in the stomach, but these are not as elaborate as those of the ruminants. Like many large mammals, hippopotami were more widely distributed in the Pleistocene than they are today.

Camel fossils occur in the Late Eocene of North America and reached great diversity there in the Miocene. They were browsers and some evolved long necks, undoubtedly having habits analogous to those of giraffes. In the Pleistocene, about a million years ago, some camels crossed into Asia by the Bering land bridge and others into South America by the Isthmus of Panama; the latter have given rise to the llama and its relatives in the southern part of that continent. Then around 12,000 to 15,000 thousand years ago camels became extinct in North America. This is another example, like the horses and the tapirs, of a group of animals becoming extinct in their evolutionary homeland and their continued existence probably depending on a handful of individuals making a journey across a land bridge to another continent. Such is the chancy nature of evolutionary survival.

Deer, with their often impressive antlers (shed annually), evolved in the Oligocene. Like the camels and ancient horses they were essentially browsers and were very diverse in the Miocene but, unlike those groups, their diversity survived the reduction in trees and shrubs as the grasslands developed. Their centre of evolution has been Eurasia, but some crossed to North America and so to the southern continent. Interestingly they have not invaded Africa south of the Atlas. Could this be due to competition from the very diverse fauna of antelopes? Related to deer are giraffes, with their special adaptation for browsing trees; they survive well in Africa and were also found in India until late in the Pleistocene.

The most successful group of large mammals today is undoubtedly the bovids—cattle, sheep, goats, and antelopes in all their diversity. They are also the last group to have evolved. There is some fossil evidence that this occurred in the Miocene; the pronghorn of North America is the only representative of a group closely allied to the ancestral stem. The main evolution of the bovids has however occurred in the last four million years or so, associated perhaps with the last temperature peak (see FIG. 10.1). Originally centered in northern Eurasia, they invaded southern Eurasia and Africa where they have diversified enormously. The bison established itself in North America and, with the pronghorn, has constituted the ungulate fauna of the plains—what they lacked in diversity they made up for in numbers. All bovids are characterized by the possession of horns covering bony outgrowths of the skull; but unlike the antlers of deer, once these are grown they are retained for life.

In addition to harbouring these large animals many types of grassland—indeed all types of habitat—are occupied by smaller mammals, more particularly the rabbit group (including hares and conies) and the rodents. Their fossil record has not been well studied but the groups were probably distinct in the Palaeocene. Well supplied with symbiotic micro-organisms in their hind guts, they have a great ability to gnaw their food with their strong teeth, the front teeth growing continuously. They are almost entirely herbivores, often specializing in particular diets in which grass and grass seeds commonly figure. Though the South American capybara stands over half a metre at the shoulder, most of these animals are small; commonly the length of the head and body is in the region of between five and ten centimetres, and the animal frequently has a tail with a similar dimension. Their small size allows most of them to lead a hidden life in the region below the lower leaves of plants and above the more compacted layers of soil; they are frequently nocturnal—dwellers in darkness, constructing runs amongst leaf litter and fallen logs. The larger species like the prairie dog and rabbit forage above ground, but have burrows to which they make a dash when danger threatens. Being vulnerable to a wide range of predators these cryptic habits are important for survival, but rabbits and rodents have also evolved very high reproduction rates; hence the expression 'breeding like rabbits'. This can lead to the rapid build up of huge populations resulting, for instance, in plagues of voles and mice and outbreaks of lemmings, when large numbers migrate seeking pastures new.

All the mammals discussed so far are placental mammals (eutherians), retaining the young in the womb where it is nourished from the placenta through the umbilical cord. There is another major group, the marsupials, in which the embryo leaves the interior of the mother at an early stage and generally crawls into a pouch on the front of the body enclosing the nipples to which it becomes attached and from which it feeds. It seems that both these groups arose at least in the Cretaceous when the marsupials may have been widely distributed on the disintegrating Pangaea, for fossils are known from North America. By the time the placental mammals started to spread, Australia and Antarctica had already split off from the rest of Gondwana so this island continent was barred to them. Marsupials died out elsewhere except in South America, where they co-existed with placentals, though by the mid-Cretaceous this too was an island continent and, as has been mentioned, it developed a fauna that was in many ways unique. Meanwhile in Australia the marsupials evolved, diversifying into most of the niches (ways of earning a living) occupied by placentals, giving many remarkable instances of convergence (whereby unrelated organisms come to have the same features because they 'do the same job'). The open forests and grasslands became populated by herds of fast moving herbivorous kangaroos and the smaller wallabies (FIG.10.8), behaving and living very much like the ungulates; more diverse than those of North America, but less so than those of Africa. Others filled the roles of rabbits and rodents, their populations suffering great reductions when European settlers introduced rabbits and rats which competed, directly or indirectly.

All these herbivores are the second link in the grassland food chain, and as has often been mentioned they are the food for further links, the preda-

FIG. 10.8 The Western brush wallaby—a marsupial of open scrubland in south-western Australia (after Ride, 1970)

tors. Most of these belong to one of two large superfamilies of mammals—the cats and the dogs. Their evolutionary history is equally as fascinating as that of their prey, but space prevents its description here. Many dogs, like the hunting dogs of the African Savannah, work in packs; operating together they can kill prey that they could not overcome alone (a strategy also adopted by animals as diverse as killer whales and ants). Most cats in contrast operate singly, stealthily approaching a prey to launch a surprise attack. However, two members of the cat group, lions and hyenas, often work as a pride or clan and may adopt quite elaborate collective ambush tactics. Some predatory marsupials evolved in Australia; they were not very diverse and the larger carnivores are now extinct. Their superficial resemblance to cats and dogs was remarkable.

On having a trunk

One group of terrestrial animals has evolved to a size large enough for its adults to be mostly beyond the grasp of predators: the elephants, whose characteristic feature is the trunk which gives them their scientific name—

the Proboscidea (FIG. 10.9). As we have already seen when discussing the challenges of life on land, weight goes up by the cube, while the cross-section of a bone is related to size by the square. Large animals therefore have to evolve to keep their legs straight under their bodies, so that the weight is taken down the length of the bones and not across them. In full-sized elephants the femur itself is slightly twisted to achieve this with the consequence that elephants walk with straight legs and they cannot gallop.

Although sauropod dinosaurs somehow managed with small heads, warm-blooded animals that feed in bulk on small items of food need to keep their mouth somewhat in proportion to their body weight so that they can take in enough food; thus large animals (baleen whales, elephants, hippopotami) seem to have particularly large mouths. One way of taking an ample mouthful is to sweep the food up and stuff it in. This is basically what the elephant does with its trunk, which is essentially an enlarged upper lip and nose. The nostrils are extended to the tip, so food can be checked for edibility before being swiftly transferred to the capacious mouth. Elephants are hind-gut fermenters, which means they do not get as much goodness out of their food as those animals that chew the cud; this combined with their large body weight means that they may need to spend 15 to 19 hours a day feeding.

A feature of the elephants from the Miocene onwards has been the development of front teeth into two or four large tusks of often spectacular size and shape (FIG. 10.9); otherwise their general appearance seems to have been very similar. The tusks have fulfilled a number of functions: in food gathering, such as grubbing up plants or stripping bark, in defence, and in male rivalry. Altogether around a hundred and sixty different species of elephant have been recognized, most of which occurred in the Miocene. Only the family to which our living elephants—the Asian and the slightly larger African—belong, evolved after the Miocene. Other members of this family included the mammoths (*Mammutus*), the dwarf species of which existed until about 4000 years ago. Representatives of several other groups lived on into the Pleistocene, but with much decreased diversity, so that the final extinction seems more like a *coup de grâce* than the cutting off of a group in its prime.

Elephants (proboscoideans) had their origins in Africa and soon (in geological terms!) spread to Asia; after colonizing every continent except Australia and Antarctica, they are now restricted to their evolutionary homelands—by contrast with horses, camels, and tapirs. They have lived in

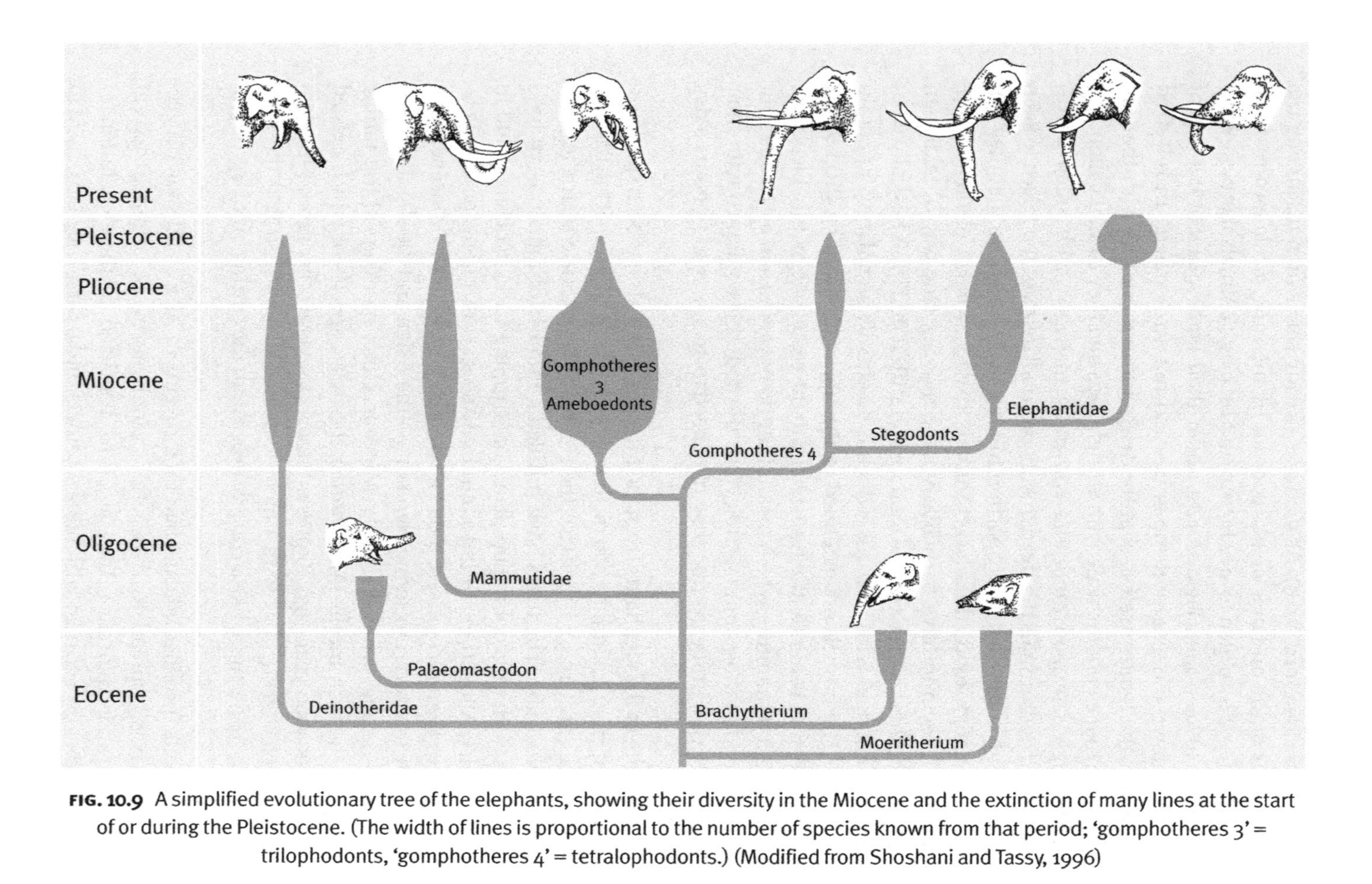

FIG. 10.9 A simplified evolutionary tree of the elephants, showing their diversity in the Miocene and the extinction of many lines at the start of or during the Pleistocene. (The width of lines is proportional to the number of species known from that period; 'gomphotheres 3' = trilophodonts, 'gomphotheres 4' = tetralophodonts.) (Modified from Shoshani and Tassy, 1996)

a wide variety of habitats, but probably always needed access to free water in some form. Crossing the Bering Straits land bridge at various times from the Miocene onwards, gomphotheres (FIG. 10.9) penetrated through to South America. The group includes the 'shovel-tuskers', which had two pairs of tusks, the lower ones flattened and blade like. It was suggested that they were specialized feeders and that they used these for scooping up aquatic vegetation; but a recent study, by David Lambert of the University of Florida, of the wear pattern on the tusks, indicates that some species used them for cutting vegetation, e.g. small branches off a tree. Others were generalists, using their upper and lower tusks for gathering plant material in a variety of ways. Some species of gomphothere lived in the rainforests of Central and South America where they probably played an important role in dispersing the seeds of certain trees. Elephants in general, like other hind gut fermenters, can live on large quantities of high-fibre food but, again like others with the same digestive system, the quality of their nutritive intake is much improved if fruit is included (the penchant of horses for apples is a reflection of this). It has been noted that some of these tropical trees now seem ill-adapted as their fruits drop on the ground and mostly rot. Are they missing their co-evolved seed dispersers—the gomphotheres? Today's elephants often shake a fruiting tree and then, with their trunks, sweep up the fallen fruit.

The mastodonts (mammutids), with long curved tusks and covered with brown-red hair, crossed to North America in the Pliocene (3.5 Mya) and spread throughout North America, becoming particularly numerous in the area around the Great Lakes. Many remains have been found, the youngest from merely 10,400 years ago. From the analysis of pollen around the teeth, and plant material in the region of the gut, a good idea can be gained of their habitat and diet. They lived in spruce woodlands, where there would be also small quantities of pine, birch, and other trees. They appear to have browsed these trees and occasionally grazed. One species seems to have selectively eaten alder, a plant with the ability to fix nitrogen and perhaps having more nutritious foliage.

The third and latest group of proboscids to invade North America were the mammoths, which belong to the most recently evolved family, the Elephantidae and not to the Mammutidae, but have the confusing scientific name *Mammuthus*! They arrived about two million years ago and spread throughout the continent, as far south as Mexico. Several different species were present, possibly representing a series of immigrants, for some were also found in Eurasia and Africa. The imperial mammoth stood up to four metres at the shoulder and probably weighed nearly ten tons; together with several other species it frequented deciduous woodlands; these must have had open areas, for samples of dung found in caves show that they fed particularly on grasses and sedges. The woolly mammoth, as its name suggests, had a dense covering of black hair and a thick fat layer to keep out the cold because it lived on the tundra steppe, in North America, Europe and Asia. During the Ice Ages this vegetation was far south of its present arctic distribution (see FIG. 10.15), and the mammoths moved with it. Occasionally specimens are found frozen in crevasses in the arctic ice and examination of these show that they fed on grasses, other arctic plants, and dwarf trees, but one wonders how they gathered enough to meet their needs, particularly when—between Ice Ages— the tundra was within the Arctic circle and in the winter there would be little or no daylight. But the frozen specimens show that full-sized mammoths flourished right up to 5500 years ago.

Elephants spend quite a lot of time in water and they are good swimmers; they colonized a number of islands: the Channel Islands off California, numerous islands in the Mediterranean and in south-east Asia, and Wrangel Island in the Arctic Ocean. What is very interesting is that in these situations they have evolved to become much smaller, these pygmy elephants sometimes standing less than a metre high when full grown (FIG.10.10). Several different groups of elephants have evolved dwarf species: stegodontids in Indonesia and Japan, mammoths in the Arctic, California, and Sardinia and true elephants in Malta, Sicily, and various Greek Islands. Why did this occur? On these islands there was a limited amount of food and, if the population built up, there would have been selection for small individuals that could reproduce on a low level of nutrition. On the mainland there were large predators like lions and tigers that prey on young elephants if they stray from the protection of their mother, which would also be the fate of a small mature elephant. On the islands there were no large predators, hence this selective pressure would be relaxed. We can also conclude from this that large size is maintained on the mainland at least in part by its value as a defence against predators. It is interesting that these pygmy elephants do not have the twisted femur that

FIG. 10.10 Comparative sizes of an Indian elephant and an extinct dwarf elephant

has been interpreted as an adaptation to get the legs centrally under their great weight. Pygmy elephants, it may be supposed, were altogether more frisky. Unfortunately we cannot be sure because they are all now extinct (probably from hunting by man), although the dwarf mammoths on Wrangle Island outlasted all other mammoths, living until 4000 years ago.

Life in the soil

A complete contrast to the world of the elephants is that of the many animals that frequent the soil. Some just make burrows in which to shelter and sometimes to store food. Many rodents do this as do some marsupials and the hyrax, a distant relation of the elephants (though a mere quarter of a metre high). Also in the soil are a vast array of small invertebrates either feeding on plant roots or plant debris or preying on those that do. Throughout the world, where there is some moisture, earthworms are a major component of this hidden life. Although in Australia they may be over two metres in length, they are frequently in the range 5 to 12 centimetres. These worms and more chunky beetle and moth grubs make a good mouthful for a small mammal. As well as roots, bulbs and tubers sit in the upper layer of the soil and can be an all-season source of food for a herbivore. In every continent at least one group of mammals has adopted the subterranean lifestyle and exploits one or other of these food sources. Although they belong to quite different groups these 'moles' have often evolved the same solutions and look very similar (FIG. 10.11)—another example of evolutionary convergence.

Living in a dark world, subterranean mammals can make little use of the sense of sight and are either blind or with very reduced eyes, probably only capable of distinguishing light from dark. The senses of smell, touch, and hearing are acute. Spending much of their lives digging through the soil, their feet, especially the front pair, have evolved to become shovel-like with strong claws, while the tail has been shortened or virtually lost.

Judging from the fossil record the subterranean habit first evolved in the late Eocene, a period of cooling and dryness, but was increasingly adopted through the Miocene and was perhaps especially associated with the spread of open grasslands and savannahs. The best-known moles belong to the family Talpidae and can be called true moles, their molehills often being features of dampish grasslands in Eurasia and North America. Their velvety silver-black fur, ideal for slipping through a damp burrow, was formerly much

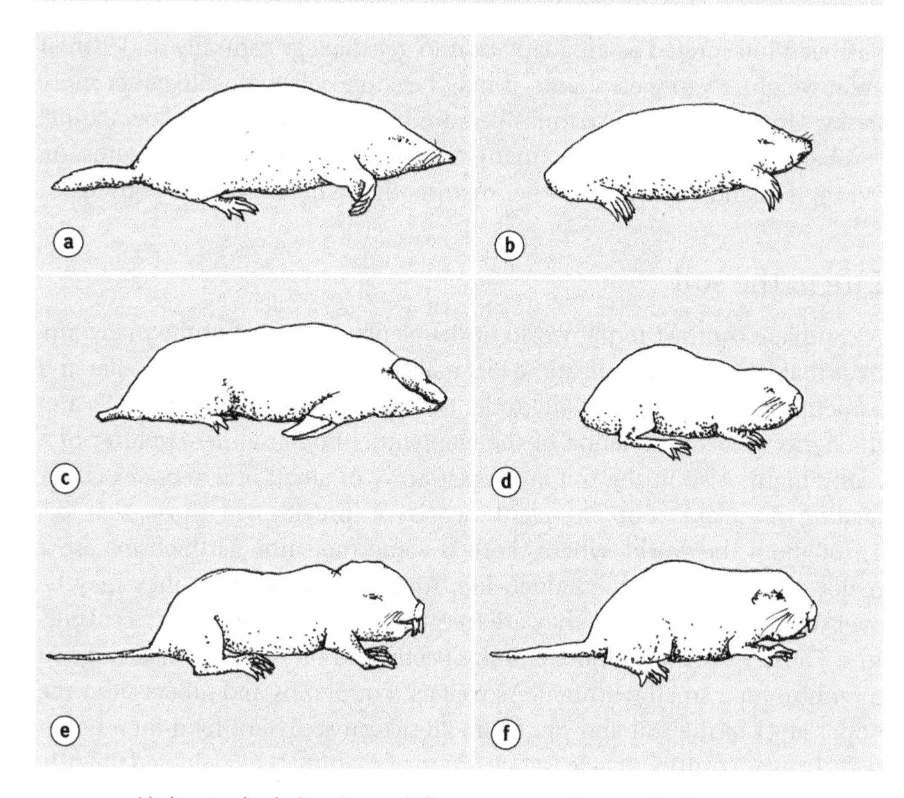

FIG. 10.11 Various moles belonging to different groups of mammals showing convergence in form associated with the subterranean lifestyle:
(a) Eurasian mole (Insectivore, Talpidae); (b) Golden mole (Insectivore, Chrysochloridae);
(c) Marsupial mole (Notoryctidae); (d) Blind mole rat (Rodentia, Spalacinae);
(e) African mole rat (Rodentia, Bathyergidae); (f) Tuco-tuco (Rodentia, Ctenomidae)

valued for moleskin waistcoats and other garments. They feed on earthworms and, especially in the summer, on insects. Before eating a worm they draw it between their front feet so squeezing out the soil from inside it. Needing to eat about half their body weight in worms daily they spend a great deal of time patrolling their burrows seeking food items that have fallen into them. In times of plenty they may create a store of worms, biting them through the head, but not killing them. There is clearly every evolutionary reason why they should not share their burrows, each individual needing its hunting area for itself and thus, not surprisingly, they are very territorial.

Africa has two types of mole—quite unrelated. The golden moles have a diet consisting of items such as insects, earthworms, slugs, and small lizards; they may seize an animal just above their burrow and drag it below. As one would expect each mole has its own set of burrows and is intolerant of intrusion by another golden mole. However, in Africa there are other subterranean animals, African mole-rats that are herbivores, and these it will tolerate. With many vegetarians the major challenge is not finding food but harvesting it, and consequently mole-rats are tolerant of each other. There is one species, the naked mole-rat, that has the most remarkable social system of any mammal; in essence it is much like that of social insects such as ants and bees. Up to 80 individuals may live in one colony, but only one female will breed. Most of the other and smaller individuals in the colony form a worker caste—digging new burrows, finding and carrying food. It is thought that there is some hormonal discharge from the 'queen' that suppresses reproduction in the other females. If she dies another female will in a short time take her place but, as the dominant female has been known to live for ten years in captivity—an exceptionally long time for a small mammal—this is hardly a frequent opportunity.

In a different group to the African mole-rats, but also rodents, are the blind mole-rats that live in dry localities around the Black and Caspian Seas. Their eyes are completely covered by skin and functionless. They are vegetarians, but live particularly on scattered resources such as bulbs, and are usually solitary. Another totally blind mole is the marsupial mole of Australia. It is of course fundamentally different from other moles, being a marsupial and carrying its young in a pouch. Its major adaptations are, however, similar in many respects to other moles, but it has a horny shield extended from the nose over the front of the head and probably forces its body, protected by the shield, through the soil; the burrows made in this way collapse behind it, whilst those of other moles are more long-lasting.

In South America there is no animal than has evolved a totally subterranean life. The nearest are the many species of tuco-tuco that have retained functional eyes and substantial tail. They look more like voles than moles and may pull food down into the burrow. They are really more comparable to the pocket gophers of North America and to the many other rodents that make burrows, but mostly gather their food above ground. That South America does not have a 'real mole' is perhaps surprising as it is believed that grasses and, presumably, open habitats were established there earlier than elsewhere. Natural history is rich with such paradoxes and prompts the question: is the statement correct? If so, how did it happen?

Thus in the Tertiary, on every continent except South America and

Antarctica, at least one group of animals has evolved to live in this habitat—the soil. Drawn from five different groups of mammal their adaptations show many strong convergence features, as shown in FIG. 10.11.

Return to the sea

The Cretaceous extinction removed some of the large predators from the seas, sharks alone remaining. The fossil record suggests that by the mid-Eocene (20 million years later) warm-blooded animals were beginning to explore the seas; this was a period, not only of warm temperatures, but one in which the sea level was high. Parts of continents were flooded: the Obik Sea straddled what is now Russia, the Tethys Ocean was being squeezed and its floor gradually raised. In Laurasia, in particular, there would have been many shallow seas and lagoons. We may suppose that these conditions would have provided many habitats that were suitable for the amphibious lifestyle and they might have increased the chances of survival of any animals that ventured further into the sea.

Birds may have started the return to the sea by skimming the surface waters, filling what was probably the niche of some pterosaurs. They then settled on the surface and developed the ability to dive, some using their wings as well as their feet to propel themselves along. Many groups of birds, such as ducks, divers, grebes, cormorants, and auks can now do this. The evolution of adaptations to the leg to become an efficient organ for swimming meant that it was no longer ideal for locomotion on land. These birds have to remain able to spend some time on land to lay and incubate their eggs, but their walk has become a waddle. In many parts of the world sea cliffs provide nesting sites that demand the minimum of walking, are ideal for launching into flight, and are relatively free of predators; it is not surprising that they normally harbour large colonies of sea birds. Wing movements are different in swimming and in flight (FIG. 10.12). It seems that for the larger birds, with body lengths in excess of about 45 centimetres, the wing needs to evolve into a flipper and they become flightless. Under the conditions of the Northern Hemisphere this obviously has had its evolutionary costs because all flightless auks are now extinct, the last representative, the Great Auk, dying out prior to 1860. The cause of its demise seem to have been a new predator (man) and a volcanic eruption in Iceland that eliminated an important breeding colony.

In the Southern Hemisphere there are large (over 40-centimetres-long) flightless sea birds—all belong to the penguin family. The earliest fossils,

FIG. 10.12 Guillemot (or murre) (a) flying and (b) swimming (after Gaston and James, 1998)

from the start of the Eocene, have been found in New Zealand and it seems that these birds have always been restricted to their present range: Antarctica, Australia, New Zealand and the southern-most parts of South Africa and South America, with an outlier in the Galapagos. There are 17 species living today, but considerably more species are known from what must be a very incomplete fossil record; several of these were large, up to half again as tall as the emperor penguin, the largest living penguin (FIG. 10.13). Penguins as a group, and these large species in particular, declined in the mid-Miocene, concurrently with the evolution of the seals. Penguins are however well adapted to marine life, spending up to 80 per cent of their life at sea, and coming ashore only to lay and incubate their eggs and to moult. Those with long, pointed bills feed mainly on fish and they pursue these in dives up to a depth of nearly half a kilometre; those with flatter bills feed nearer the surface mostly on krill and other shrimp-like crustaceans.

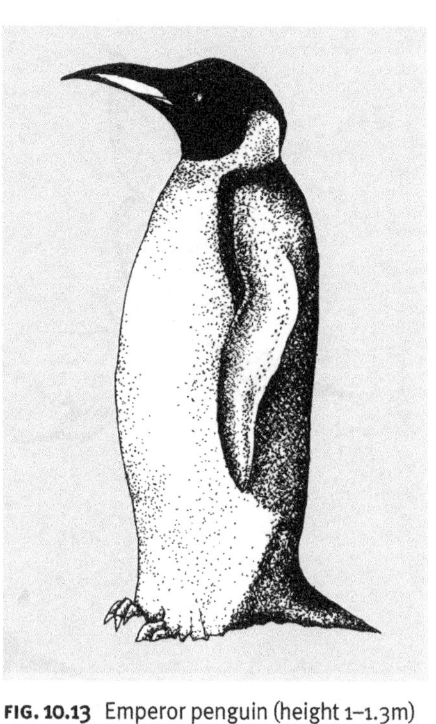

FIG. 10.13 Emperor penguin (height 1–1.3m)

The degree to which the penguins have been able to adapt to the demands of their lifestyle is well demonstrated by the almost bizarre life of the emperor penguin. It nests in the Antarctic on stable ice, often adjacent to the edge of the continent. The temperatures there may fall as low as –60°C and yet the male emperor penguin incubates the egg on his feet covering it with a special feathered flap. He will huddle with other males, all turning away from the wind; individuals change their place in the crowd so that none remains long on the windward side. Without this communal mutual assistance the males would exhaust their body reserves keeping warm, for they go without food for a total of fifteen weeks: six weeks pairing up with a mate and nine incubating the egg.

When the chick hatches towards the end of the winter, the female may have returned but, if she has not, the male will feed the chick on a protein-rich secretion from the fore gut. When relieved by the female—and she finds him from among thousands of other males, some of which will have been her mates in earlier years—the male will make his way to the sea to feed and then return to feed the chick and replace the female. They will take turns for about six weeks; each journey over the ice may be around 100 kilometres and on reaching the sea they will swim for an average of about 500 kilometres searching for food and capturing their prey, principally fish. When the chick is well covered with down and about six weeks old, both parents leave and it will huddle together with hundreds of neighbouring chicks in what is appropriately referred to as a crèche. Each parent will return about ten times, having travelled overland and at sea for a total of some hundreds of kilometres, and find its chick amongst many hundreds of others, and feed it. By now the summer has arrived and the pack ice will be melting so the parents' journeys to the open sea will be shorter and feeding visits more frequent, matching the chick's growing appetite. If the sea ice melts too quickly the ice on which the chicks are resting may break up before they are ready to swim and there is a danger that this may occur with increased frequency due to global warming.

The adults stop feeding the young, which start to lose their grey down and after a while begin to make their way to the open water, by now—in mid summer— a distance of a few to some 60 kilometres. They are ready to take to the sea in their new plumage and, with luck, killer whales or leopard seals will not be waiting for them. If they survive they will return to the colony in about four years and breed; having reached that stage of life their chances of survival are good and 95 per cent make it to the next year. They are clearly well suited to both life at sea and the stresses of parenthood and it would seem that some live for 50 years. However recruitment to the breeding population is at a low rate; on an average about three-quarters of the eggs hatch, but the really heavy mortality occurs in the immature stage: four-fifths perish between the time the parents leave the fledglings and when they return to the colony. Learning to be an emperor penguin is clearly a hazardous process.

One can but marvel at the remarkable physiological and behavioural adaptations that make the emperor penguin such an adept inhabitant of its environment. Every organism is a prisoner of its evolutionary history and, for a bird, the emperor penguin has gone a long way to recapture the adaptations of fish. But in such a cold environment its warm-bloodedness gives it an advantage over fish: it is able to respond more quickly to stimuli and thus hunt efficiently. On the other hand, it has not been able to shed its need to return to land to breed. It puts a great deal of time and energy into the production of a small number of offspring in a lifetime. Its existence as a species is sustainable only because of the low mortality of the mature adults. Species that are well adapted to particular and relatively unchanging environments often have these strategies; they maintain their population near the maximum for the habitat—the carrying capacity. Ecologists call them 'K-selected', K being the symbol used for carrying capacity. They contrast with '*r*-selected' species that have evolved to have a high rate of reproduction that best allows them to exploit an environment whose carrying capacity is always changing; hit hard by mortality, their populations have the productive capacity to spring back—one thinks of rats and mice for example. The K-species have low rates of reproduction, so if the habitat suddenly changes and mortality greatly increases, they cannot recover and become extinct.

One can see how the emperor penguin depends on the Antarctic environment remaining much as it is. Changes such as less or more ice, a new predator (especially one on land where they are so vulnerable), a fatal disease of crowded birds, or fewer fish—so that the parent has to go some hundreds of kilometres further to find food for the chick which starves while it waits—could all lead to its decline. Since huddling together is necessary for the survival of the incubating males and of the fledgling chicks, a small population is a doomed one.

It is the members of two groups of marine mammals that are the main enemies of penguins in the seas and, of course, they have also adapted in their return to the sea. The seals (Pinnipedia), like the penguins, have retained their dependence on dry land for reproduction and have retained their limbs. Like auks and penguins they too walk awkwardly on land but, modified into flippers, the limbs provide much of the power for swimming although the body is also flexed. The whales, dolphins, and porpoises (Cetacea) have become entirely marine and, although the front limbs are retained as flippers, their function has more to do with steering than powering swimming; this is done by the tail that has developed two large flukes, folds of the skin strengthened by connective tissue.

The Pinnipedia have the shorter evolutionary history and are not found until the Miocene. There are three groups, sea-lions, walruses, and seals; most evidence suggests that they arose from the ancestors of bears. They are all carnivorous, mostly feeding on fish, but given the opportunity will also take sea birds; the walrus—as Alice in Wonderland learnt—specializes in oysters and other shellfish on the sea bottom.

Whales have a longer marine history than seals, the first fossils being found about the same time as those of penguins, in the early Eocene, when they took to the sea around the eastern margins of the ancient and shrinking Tethys Ocean; by the end of the Eocene they had become widely distributed in the oceans and seas of the world. Derived from ungulates, perhaps not far from the origins of hippopotami—and with a similar lifestyle—their hind legs had by the mid-Eocene lost their locomotary function, therefore they could not come on land to give birth. This is shown by a fossil whale, *Basilosaurus*, from this period in which the hind limb is rudimentary, probably functioning only as a copulatory guide. Thus within a period of probably less than ten million years whales had made the major evolutionary, but seemingly difficult, step of giving birth to air-breathing young in water. This freed their evolution completely from any residual need to function on dry land. However reproduction does seem to have placed one requirement on many of the large species: they return, from polar regions, to warm and shallow waters to give birth and, often, to subsequently pair. This minimizes the energy cost to the young in the first few weeks of life.

Whales and their smaller relatives, porpoises and dolphins, have become highly streamlined with surface structures that reduce drag. Even the hitherto puzzling lumps on the fins and heads of humpback whales have been shown to reduce drag by 10 per cent and increase lift by five per cent. The effect of these adaptations and of the water buoying up their weight is that they can swim at speed for long periods of time. Dolphins can move at 40 kilometres per hour for hours; land animals can achieve speeds of almost twice this, but can only maintain them for a matter of minutes.

There are two groups of whales with different food-capturing mechanisms: the toothed whales and the baleen whales. The toothed whales, which include dolphins and porpoises (FIG. 10.14(a)), are carnivores, mostly taking fish, sea birds, seals and other whales, though some are especially partial to squid. The killer whale is the top predator in the seas, not simply because of its large size (5–6 metres in length), but particularly because it hunts in packs and these will attack much larger whales. The sperm whale, the hero of the novel *Moby Dick*, is the largest toothed whale (17 metres), but is not a rapacious carnivore, feeding almost entirely on squids, including the giant squid of the deep ocean. To hunt this prey the whales dive deeply, commonly between 500 and 800 metres, but sometimes up to a kilometre, and they can remain underwater for anytime up to an hour.

Baleen whales (FIG. 10.14(b)) are large-scale filter feeders; they trap their food—krill and other shrimp-like crustacea—in their baleen. This is a structure made of keratin (as are our finger nails), it hangs down like a curtain and, as mouthfuls of water pass out through it, the food organisms are trapped and then swept off to the throat by the tongue. There are three basic feeding methods. The skimmers, such as the right whales, swim slowly through the water with their mouth open, the water entering at the front and passing out through the baleen at the sides. The rorquals, which include the largest of all whales, and of all animals that have ever lived—the blue whale—are gulpers taking in huge gulps of water into the mouth, the underside of which is pleated allowing great expansion; this water is then forced out through the baleen. Lastly the grey whale, a grubber, scoops up material from the muddy bottom of the sea, filtering out worms and other prey as the water and fine particles pass out through the baleen.

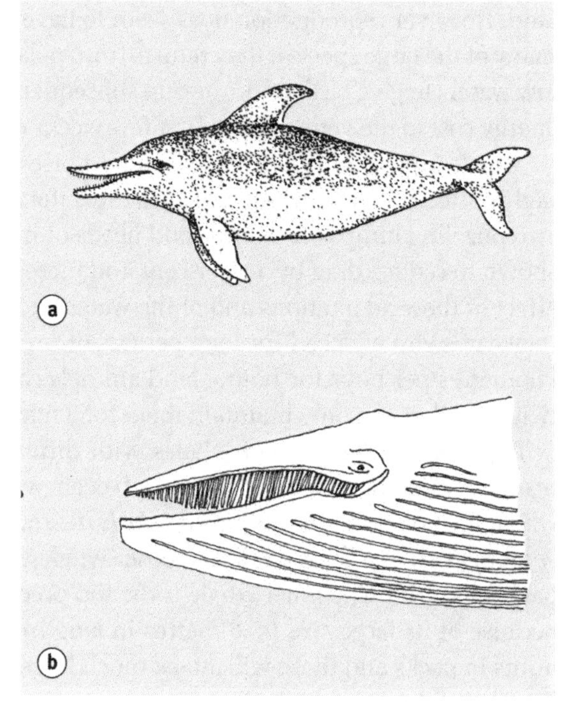

FIG. 10.14 Whales: (a) dolphin (a small toothed whale); (b) diagram of the head of a baleen whale (after Fraser,1976)

The larger whales have, like most large animals, few enemies; they are long lived (the bowhead over a hundred years) and reproduce slowly. They fit the definition of K-selected organisms explained above. A new cause of mortality is likely to drive them to extinction, and hunting by man has posed just such a threat. Even those whales that do not live in obvious groups seem to communicate over significant distances. The young have a long period, extending for at least 13 years, when they are learning from the adults. The learning process probably continues; navigating the oceans of the world, the older individuals may have an important role with their accumulated experience. These social links make whale populations more vulnerable than numbers alone might suggest and therefore the international ban on hunting was necessary and should be maintained. We should also avoid accidental deaths from other fishing activities. The populations of many species, though not all, are showing signs of recovery. The northern right whale looks doomed and the blue whale population is still not building up to its previous levels.

There is one other group of mammals that took to the seas in the Eocene, the sea cows, dugongs or manatees (Sirenia) which are related to the elephants. It is interesting that all the major groups of large-bodied terrestrial placental mammals, the carnivores, the ungulates, and the elephants, had branches that returned to the sea in the Tertiary, but not the smaller-bodied rodents or primates—unless we accept the theory that humans were marine primates (p. 226). The sea cows, as their name suggests, are herbivores. Although there are now only four species, they are found, very locally, in shallow coastal waters in almost all parts of the tropics and subtropics. They are supposed to be the origin of the sailor's myth of the mermaid, but imaginations must have been greatly stretched, for they are large ungainly animals weighing up to a ton and a half! Like other large marine mammals their populations have suffered from depredations by man; in 1742 the six-ton Steller's sea cow was discovered in the shallow parts of the Bering Sea, however by 1769 it had been hunted to extinction.

The quaternary period

This period, covering approximately the last 1.64 million years, has seen the evolution of some apes, the hominins (see footnote on p. 216), which developed an increasing ability to modify the environment and to affect the lives of other organisms. But they, and all other organisms, were affected by the many ice ages and other changes in climate, such as periods of dryness in Africa, during the first 1.63 million years of this period—the Pleistocene.

Knowledge of the past climate is gained from several pieces of evidence. The organisms found in deposits are generally those alive today and so we know the conditions they require; for example if the species found in a deposit are now found in Spain, then we know that the deposit was laid down in a warm and dry environment. The next step is to find the age of the deposit in which they were found. This can be done by radio-carbon dating which is based on the very small amount of the radioactive carbon isotope, ^{14}C, that is present in the atmosphere. Animals and plants take this into their bodies, in their exchange of carbon dioxide, but when they die this carbon is held in their tissues and slowly decays—half of it in 5730 years. Therefore by measuring the amount of this isotope in a fossil it can be aged provided it is younger than some 70,000 years. A measure of actual temperature in sea water can be gained from the ratio of two isotopes of oxygen, ^{16}O and ^{18}O, as the latter, which is heavier, is relatively more abundant if the sea water is cold. Shellfish take these isotopes of oxygen into their shells, in calcium carbonate, in the same ratio as that of their environment and this ratio will be maintained in the dead shell; this can be retrieved in deep sea cores.

Information is also gathered from the nature of the deposits (such as wind-blown soils indicating dry conditions), from the concentration of carbon dioxide in air bubbles trapped in the polar ice sheets (this will be lower if it is trapped at a time of widespread glaciation), and from the growth rings of trees and corals. The new wood in the trunk of a tree forms a distinct ring each year, and will be narrow if the tree is stressed or broad if growing conditions are good. The remarkable bristle-cone pine of the Rockies, which can live 8000 years, provides a very precise record for that period. Somewhat comparable are the bands of different densities in the skeletons of stony corals that appear to reflect differences in water temperature and illumination.

It seems that there have been many glacial cycles during the Pleistocene; the number of cycles is very dependent on the extent of the cooling that is considered sufficient for a time to be counted as a glacial episode; a total of about 15 glaciations would be a reasonable compromise. Many studies have been made to ascertain their cause. The major factor seems to have been cyclic variations in the pattern of incoming solar radiation due to changes arising from the nature of the earth's orbit: these are termed Milankovich cycles and can be calculated from astronomical information. However, they are not the whole story. It has been found that the variations in sea temperature (as determined from deep sea cores) over the last 350,000 years is best explained by a combination of Milankovich cycles and the carbon dioxide level—the greenhouse and reverse-greenhouse effects.

Evidence of ocean temperature from deep sea cores suggests that from about 1.3 million years ago there was a phase in which the cold pulses were quite short and lasted about 30,000 years, but about 800,000 years ago the cold periods became longer. Most studies in the northern hemisphere suggest that there were four major glacial periods in the subsequent 700,000 years. As Europe, Asia, and North America had been moving north, there were now large land masses in the proximity of the North Pole. Hence enormous ice sheets formed in the northern hemisphere during the cold phases and where these were on the land (shelf ice), sea levels would have fallen. Considerable quantities of water would have been locked up in this ice and so rainfall would generally be low. These ice sheets would not only have blocked the flow of rivers, particularly those in Russia that would otherwise have flowed north into the Arctic Ocean, but would also have changed the direction of the ocean currents. Both these events would have direct and often quite rapid effects on regional climate.

The Middle Pleistocene, which is generally taken as commencing about 700,000 years ago and lasting until 127,000 years ago, was marked by three main glacial periods in the northern hemisphere. The fourth and last glaciation was in the Upper Pleistocene, a period that extends from 127,000 years ago to the end of the glaciation, some 10,000 years ago. These glacial periods and the intervening interglacials have been given different names in different regions and there is still some uncertainty as to how closely they can be correlated. Because these terms often occur in other writings about the Quaternary they are listed here, the interglacial periods are in italics (TABLE 10.1)

At the outset of the last glacial period the sea temperatures were not very different from today but between 120,000 and 80,000 years ago gigantic ice sheets had started to develop in the northern hemisphere. These reached their maximum about 18,000 years ago when ice covered most of Canada and northern USA, a large part of north-west Europe, with another sheet over the Alps (FIG. 10.15), and at least part of Siberia. It is thought that there was a less widespread glaciation in the southern hemisphere where, as throughout the Quaternary, Antarctica was covered with shelf ice. The ice ages of the northern hemisphere were, it seems, marked by more extensive sea ice. In places, the northern ice sheet was probably as much as three kilometres thick and, with this amount of water taken out of circulation, it is perhaps not surprising that the sea level was at least 120 metres less than today. Only the southernmost parts of Europe were covered with trees, largely coniferous; much of the middle region was tundra or forest tundra, the haunt of the mammoth and woolly rhinoceros.

Table 10.1 Names of the glacial periods and intervening interglacials (italics) for different regions

	Rhine Estuary	**Britain**	**Alpine**	**North America**
Upper Pleistocene	Weichsellan	Devensian	Würm	Wisconsin
	Eemian	*Ipswichian*	*Riss-Würm*	*Sangeman*
Middle Pleistocene	Saalian	Wolstonian	Riss	Illinoian
	Holsteinian	*Hoxnian*	*Gt.Interglacial*	*Yarmouth*
	Elsterian	Anglian	Mindel	Kansan
	Cromerian	*Cromerian*	*Günz-Mindel*	*Aftonian*
	Menapian	Beestonian	Günz	Nebraskan

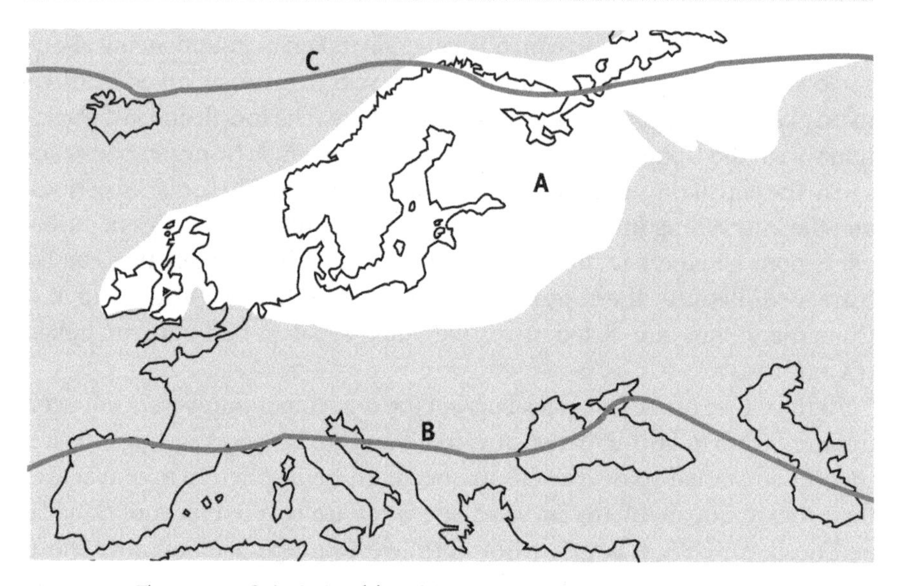

FIG. 10.15 The extent of glaciation (A) at the maximum of the last ice age and the position of the northern limit of timber, then (B) and now (C) (after Goudie, 1983)

The last ice age ended relatively quickly. The warming started around 17,000 years ago and continued until 11,000 years ago when there was a reversal of about 2°C lasting for a thousand years (the 'Younger Dryas period'). When melting recommenced it proceeded quickly; the rate of discharge from the Laurentide (Canadian) ice sheet between 10,000 years ago and 8000 years ago has been described as 'almost catastrophic'. In Britain, a sudden change came about when the North Sea unfroze and ocean currents started to bathe the shores once more. This demonstrates how climatic change is not always gradual—there may be a step change.

It would be wrong to think that conditions have been maintained much as they are today for the last 8000 years. There were warmer periods around 6000 years ago and again 1250 to 700 years ago (750–1300 AD), at the peak of which wine was widely grown in England. There have also been three colder periods, around 5000, 3000 and 200 years ago. Now the global climate is getting warmer, but there is a new factor at work— an able, powerful, and populous ape.

CHAPTER 11

The Evolution of the Able Ape

20 Mya–30,000 years ago

THAT able ape, mankind (*Homo sapiens*), belongs to the primate group of mammals. This probably evolved in the Cretaceous from insectivorous mammals, not far distant from today's shrews. They have a number of characteristics: a conspicuous one is the retention of the primitive five digits, with the thumb and big toe becoming more mobile and independent of the other digits; and as the nails are flat, rather than modified into claws, the tips of the fingers and toes can evolve a good sense of touch. In contrast the sense of smell is reduced, the snout being shortened. In keeping with their generally wide diets (omnivores), they have retained almost all the teeth found in primitive mammals. There has been a progressive expansion and elaboration of the brain. Comparisons made by Paul Harvey of Oxford University and Tim Clutton-Brock of Cambridge University have shown that, in proportion to their body size, fruit-eating species have larger brains than foliage eaters; fruits are more difficult to find than foliage, they are more scattered in space and time in the forest. As they put it, the brain, in parts, serves as a three-dimensional map of the environment. Here again we see the complex evolutionary interactions between animals and features of the vegetation, apparently simple factors leading eventually to large evolutionary steps. An early example, already discussed (p. 73), is that of clogging water channels by plants leading to the evolution of limbs, an adaptation that proved suitable for the invasion of land.

In the early Eocene there were at least two groups of primate: one more closely related to today's lemurs and the other to the monkeys and apes. This latter group, the omomyids, seems to have had many of the features of the modern tarsier (FIG. 11.1), a small agile nocturnal animal eating fruit and insects. Most of the fossils of these early primates have been found in North America and Europe and it has been difficult to relate them to the

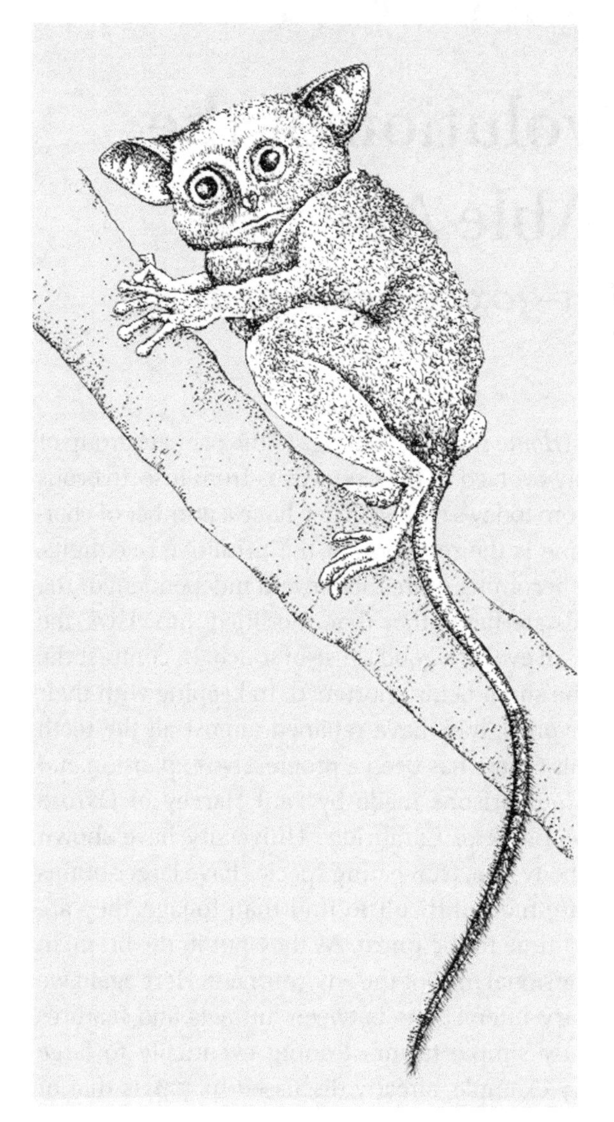

FIG. 11.1 The tarsier—a primitive primate

living members of the group, but recently fossils have been unearthed in Africa and it seems that this continent, and possibly Asia, were the centres of the early evolution of primates. The off-shoots in the other continents were evolutionary dead-ends.

There is rather more uncertainty about the evolutionary history of primates than most other groups of mammal as fossils are relatively scarce; probably the living animals were nothing like as abundant as the members of some other groups, like the ungulates, where there is a detailed fossil record.

The Miocene apes

The earliest monkey fossil has been found in the late Eocene, but the fossil record for the Oligocene is sparse and it would appear that the first hominoids—that is, the ancestors of today's great apes and man—separated from the baboons and related monkey groups early in the Miocene. Primates are essentially a tropical and sub-tropical group, adapted to live in warm environments. There are a few exceptions, notably humans and macaques, and these have special ways of keeping warm in cooler conditions. The Japanese macaque often sits in hot springs, while the rhesus macaque in the Himalayas lives around temples and is 'pampered' by the monks. It therefore seems likely that the cool climate of much of the Oligocene would have restricted the distribution of primates. In contrast, the heat of the mid-Miocene would have favoured them and there was a wide diversity of early apes at that period.

The first radiation appears to have been in East Africa where, in the early Miocene, there were several species that can be loosely grouped as *Proconsul*. They seem to have been forest dwellers spending much of their time in trees; the structure of their thumbs was such that it could enable the hand to grip—an important step in hominoid evolution. Towards the middle of the period, some 19 Mya, tectonic movements had brought Africa and Arabia against Eurasia so that animals could move between these lands. A few million years passed before the apes made this journey, but between 16 and 11 Mya there were many species living in Europe and Asia, as well as Africa. This corresponded with the warmest period in the Miocene, which was the warmest in the recent history of the earth (see FIG. 10.1). These 'middle-Miocene apes' included *Dryopithecus* that was found in south-west and central Europe, *Kenyapithecus* and *Afropithecus* in east Africa, *Graecopithecus* in Greece and *Sivapithecus* in Asia. There was a split towards the end of this time between the line that evolved to the orang-utan, represented by *Sivapithecus*, and that which included the ancestors of the gorilla, the chimpanzees, and man.

In the fossil record from east Asia a further genus, *Gigantopithecus*, appears towards the end of the Miocene. Most experts place them in the orang-utan group; *G.blacki* that occurred until at least 475,000 years ago was the largest primate ever known, weighing—it is estimated, 200 kg. They appeared to have co-occurred with *Homo erectus* and their teeth are sometimes offered for sale in Chinese drug stores. It is possible that discoveries of their remains have contributed to the legend of the abominable snowman or yeti.

The dawn of the hominids[1]

Unfortunately the fossil record from 10 to 4 Mya is sparse, even by primate standards. A fossil from 10 Mya found in Greece in 1991 has been named *Ouranopithecus* and may represent the ancestral line of the African apes and humans—the hominids. The next oldest fossil on this line, found as recently as 2000, dates from 6 Mya; it was found in the Tugen Hills, Kenya, informally called 'millennium man', it has recently been given the scientific name *Orrorin tugenensis*. This hominid fossil is particularly interesting as it lived around the time that the ancestral lines leading to the gorilla, the two chimpanzees and humans had diverged (FIG. 11.2). The evidence to determine the timing of this division has been obtained from biochemical studies of the living species: how different the key molecules are and how long would these differences take to accumulate, that is, using the molecular clock.

When in the 1963 Morris Goodman of Wayne State University suggested that humans were very much closer to the African apes than these were to the orang-utan and gibbon, many people's preconceptions of the separation of humans from apes were challenged!

This work was based on immunological studies of blood proteins. Since then there has been much further research, based particularly on the detailed structure of the DNA of both the nucleus and the mitochondria, all of which has confirmed Goodman's original conclusions. The chimpanzees and humans have around 98 per cent of their DNA in common, and both share less with the gorilla; but in many other respects, such as knuckle-walking, the African apes are very close to each other. The detailed interpretation of the information is controversial, but it has been suggested that gorillas separated from the others a very short time before the divergence of the chimpanzees and the hominins. The date for this three-way divergence varies according to the type of evidence and interpretations, but it is generally thought to be between five and seven million years ago.

[1] 'Hominoid' refers to the largest group—the apes, humans, and related fossil species; 'hominid' to the African apes and humans and related fossil species and 'hominins' only to humans and fossil species evolved after the separation from apes—those in FIG.11.3.

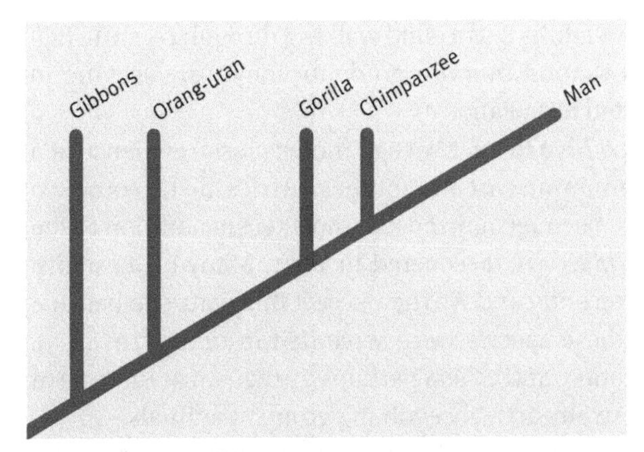

FIG. 11.2 The relationships of living apes and humans

A big evolutionary step was the development of bipedalism, that is walking upright on two legs. Although many other mammals can stand on their hind legs, they walk on all four legs, the backbone forming a bridge from which the body is hung. The African apes walk in this manner with their hands clenched—the so-called knuckle-walking. There is much debate about the evolutionary pressures that led to this change of habit, but it seems to have been associated with moving out of a forest life into more open country. The forests became more restricted in the cooler and drier climate of the late Miocene and early Pliocene (6–5 Mya) and moving on to the savannah, with lions and other large predators, the upright position would have greatly aided the keeping of a look-out. Another advantage of the additional height may have been the increase in the distance over which groups of hominins could signal friendly or antagonistic messages; in the latter case this greater field of vision could lead to the avoidance of physical contact and the consequent risk of damage from fighting. There could also have been physiological advantages: thermal stress, and hence water loss, is lower if less of the body is fully exposed to the sun, and so the need for drinking water is reduced—it has been calculated—from a minimum of 2.5 litres per day to 1.5 litres per day; furthermore, striding is an energy-efficient way of moving.

Probably all these factors, and possibly others, played a role in the evolution of bipedalism. Once the front limbs were freed from a role in locomotion the possibility arose for their use to carry things, including tools and helpless young, from place to place. That hominins did walk upright can be confirmed from their skeletons, but also from the poignant evidence of the fossil footprints found at Laetoli in Tanzania. These were made 3.5 million years ago, we believe by two adults and a child walking through recently fallen volcanic ash, changed to mud by rain; no doubt they were seeking to escape from their devastated homeland.

Over the period from 4.6 Mya to 1.6 Mya ago there is fossil evidence of a whole variety of early hominins in east and southern Africa. So far some sixteen different species have been recognized, the most recent addition to the list being *Kenyanthropus platyops*, discovered in 2001. Many of these discoveries have been made recently and we may expect the picture to become more complex. Some of these species were separated in time (FIG. 11.3), others in geographical location, and others perhaps by diet—just the type of picture one would expect in any actively evolving group of animals.

Most of the hominins living during this time can be loosely referred to as australopithecines. Their assignment to a particular genus, which gives them the first part of their scientific name, changes as more fossils are found and is a matter on which experts often disagree. The specific name, the second part, does not change (though it may be decided that a particular fossil was misidentified and should be renamed); so to follow the story in this and other books one should concentrate on the last part of the name.

Apart from the earliest species, *Aridipithecus ramidus*, that lived in Ethiopia 5.3 Mya–4.5 Mya, the other australopithecines can be broadly divided into two groups—the robust and the gracile. The former are placed in the genus *Paranthropus*; they had strong jaws and huge grinding teeth. In contrast to the generalist feeding habits of other australopithecines and *Homo*, they are thought to have been specialist vegetarians living on nuts, tubers, and other tough parts of plants. *Paranthropus aethiopicus* was found in east Africa 2.5 Mya, at the onset of a dry period in the Pliocene. It seems to have been replaced by *P. boisei* in east Africa, while two other species lived in South Africa. Some of these robust species existed until well into the Pleistocene and would have been contemporary with the early species of *Homo*. At a site in Kenya 'Nutcracker man', *P. boisei*, seems to have occurred in the same countryside as some species of *Homo*. Did they interact or did they ignore one another? We cannot know but, judging from the way that chimpanzees will turn on other primates, including their own species, and likewise one group of *Homo sapiens* on another, it seems most probable that if they came into contact they were unlikely to be friendly. Again we have no way of knowing if this had anything to do with the eventual disappearance of the robust line but, like many specialized groups *Paranthropus* proved to be an evolutionary dead end.

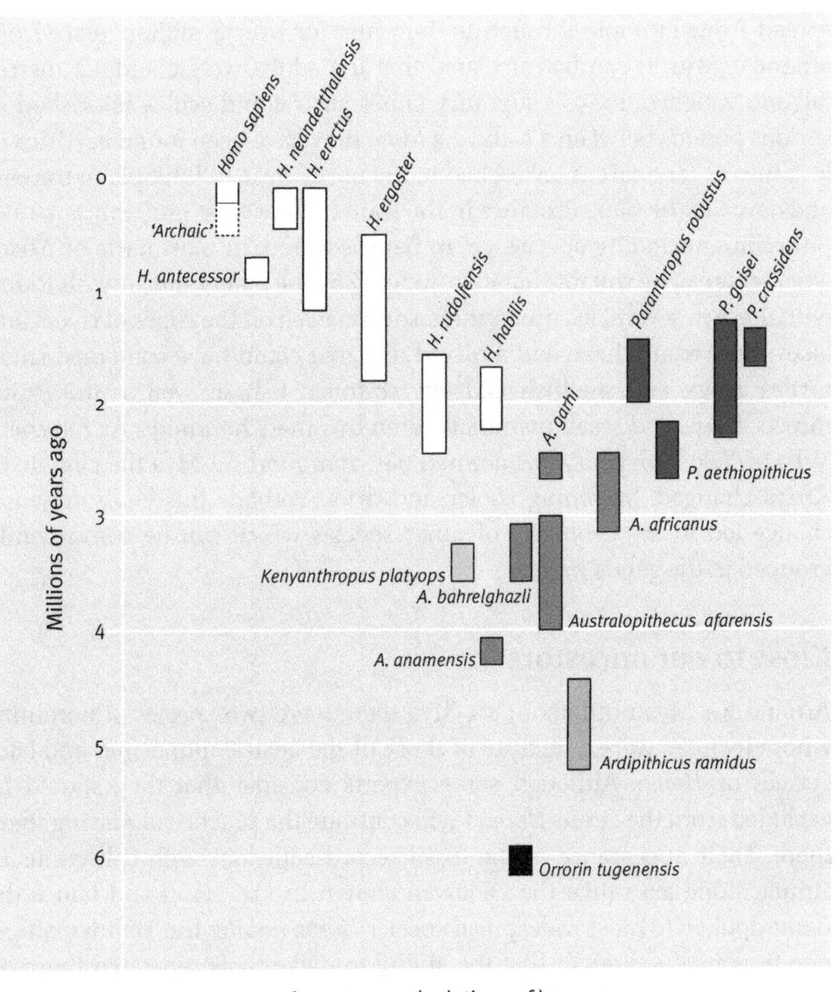

FIG. 11.3 Ancestors and relations of humans (modified and expanded from Aiello and Collard, 2001)

There are some six species of australopithecines currently recognized that may be considered as belonging to the gracile or slender group and it is likely that some at least were on or very close to the line from which *Homo* evolved. The best known is *Australopithecus afarensis*, a partial skeleton of which has been nicknamed 'Lucy'[2]; the species seems to have been wide-

[2] On the evening after the first parts of the skeleton had been found, her discoverers, Donald Johnson and Tom Gray, repeatedly played the song *Lucy in the sky with diamonds* by the Beatles on their gramophone.

spread from Ethiopia through to Tanzania for over a million years from around 4 Mya. It can be calculated that the adults were about 1.4 metres tall and weighed 40–50 kilograms. Other australopithecines flourished in various periods between 3.5 and 2.3 Mya; they have been found in Africa in localities as far apart as Lake Chad in the north-west, to Ethiopia in the east and down to the Cape Province in the south. There were probably australopithecines, including species yet to be discovered, in most parts of Africa where there were suitable habitats. Judging by the plants and animals found with them in several locations, these species lived on the edges of rivers and lakes; they would have had a mixed diet that could have contained small turtles, frogs, and shellfish and any stranded fish, as well as the fruits, insects, birds, and small mammals eaten by others hominids. As Elizabeth Vrba of Yale University has pointed out, at around 2.4 Mya the climate of Africa changed, becoming cooler and drier. Perhaps this environmental change led to the evolution of other species which can be conveniently grouped in the genus *Homo*.

Close to our ancestors

Around 2.4 Mya until about 1.5 Mya there were two species of hominins whose features were a mixture of those of the australopithecines and later species of *Homo*. Although some experts consider that they should be excluded from the genus *Homo* I will continue the practice of placing them there. Their appearance in the fossil record coincides with collections of simple stone axes (like the Oldowan shown in FIG. 11.4) and hence the name applied to most widespread species *Homo habilis*, the 'Handyman'. At one time it was thought that the ability to make tools separated those on man's evolutionary line from the African apes. However we now know that not only do chimpanzees use tools, they may also make them. For example they fashion sticks to extract termites from their nests; in captivity a bonobo (one of the two species of chimpanzee) made sharp stone flakes by throwing a stone on a marble floor, and then used these flakes to cut string and gain access to a favoured food. Chimpanzees commonly use rocks or hard pieces of wood as hammers and anvils to crack nuts, even using another stone to wedge under the anvil to make it flat. In different parts of Africa they have what would be called in human societies different 'traditions' of tool use. When chimpanzees need a tool they find one nearby and, when they have finished the task in hand, they put it aside. In this they differ from humans –their tools do not become long-term possessions. However, when they next need a tool they may search out the old one and, if necessary, carry it a short distance to the new feeding site.

The extent to which the work of the handymen was a major advance on that achieved by chimpanzees, and perhaps by australopithecines, is controversial, but two features seem to be significant. Firstly, stones worked by handymen are found up to about ten kilometres from their natural location so they must have been carried with the intention of fashioning and using them. Secondly, those worked stones found allow the interpretation that one stone was used to fashion another, chipping off flakes; where such a tool was used to make another tool it is termed a secondary tool. It has been

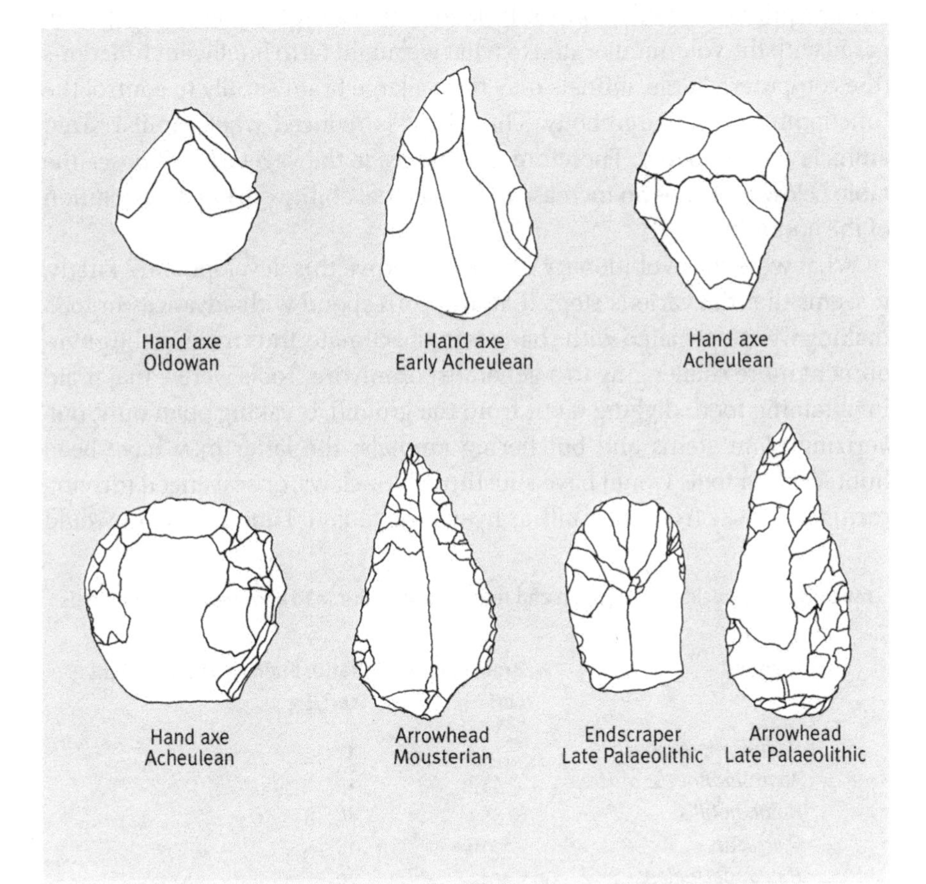

FIG. 11.4 Stone tools from various periods showing increasing complexity (after Oakley, 1975)

proposed that the use of secondary tools marks out the genus *Homo* from other hominids.

The size of the brain of the australopithecines was greater than that of even earlier hominids and another step in this increase occurred in *Homo habilis*, as can be seen from the TABLE 11.1

Though tempting, it is probably misleading to simply interpret a larger brain/body ratio as reflecting more intelligence since the body weight of a herbivore, with its need of a large hind gut for digestion, will be proportionally greater than that of an omnivore. Thus, although intelligence may be equal in the two animals, the herbivore will have the lower ratio. In animals with fundamentally different designs the brain will differ in the proportions required for body control functions (the telephone exchange) compared with the volume allocated to what we might term intelligent functions (the computer). Large animals may have a large brain simply to control the functioning of the large body. This effect is reduced when similar sized animals are compared. Therefore, leaving aside the vegetarian *P. bosei*, the table below indicates an increase in intellectual ability during the evolution of the hominins.

What were the evolutionary forces that drove this development? Firstly, it seems that the various steps that also correspond with advances in tool-making were associated with changes in the climate, thus making the environment more challenging to a generalist omnivore. Tools were a major aid in obtaining food: digging it out from the ground, breaking open nuts, pulverizing plant stems and butchering animals; the latter may have been hunted when tools would have substituted for claws, or scavenged (driving carnivores away from their kill, as hyenas do today). Time and effort would

TABLE 11.1 Increasing size of brain and ratio of brain volume to body weight in hominids

Hominid	Brain volume cm^3	Ratio: brain vol/body weight cm^3/kg
Paranthropus bosei	513	12
Australopithecus africanus	457	13
Homo habilis	552	16
H.errectus	1016	18
H.neanderthalensis	1512	20
H.sapiens	1355	26

have gone into the making of an Oldowan axe (FIG. 11.4) and it seems likely that these were carried over distances, made possible by the bipedal stance evolved in the australopithecines; although bipedal handymen also seem to have had good climbing abilities.

The evolution of a larger brain would be favoured by the practice of making tools, by walking in search of scattered food (as the savannah spread) and by having some 'possessions'. These features would also increase the amount of information the young would need to learn and hence the time that the young remained dependent on their parents. The more one generation could learn from the previous one, the more successful it would be, and the faster the evolution of living habits—that is of culture. Success would also come from the sophistication of social skills, both within the family group and in encounters with other groups. Evolutionary psychologist Richard Byrne has pointed to the evolution of social expertise, such as having the ability to gain insight into the thoughts of another, as a key driver in human evolution. We can visualize that this Machiavellian intelligence would have been important, for instance, in achieving peaceful co-operation between family groups which in turn could have led to an increased effectiveness in hunting for what was probably a valued component of diet—meat.

The Turkana Boy and the Erects

Around 1.8 Mya there was another period of global cooling and at the same time new types of tool were made, termed Acheulean, which are characterized by having both faces flaked to give a cutting edge (FIG. 11.4). They were used throughout Africa, Europe and western Asia until a few hundred thousand years ago, sometimes being as much as 20 kilometres from the site of the rocks from which they were made. The fossil record also reveals from the same period the emergence of two new species of *Homo: H. ergaster* and *H. erectus*. The best known specimen of *H.ergaster* was found near Lake Turkana, Kenya; it was of an adolescent boy—hence the informal name, and it seems that had he survived to adulthood, he would have been over six feet (1.8 m) in height. Both these species had less tree-climbing ability than Handyman, which perhaps always lived in the vicinity of trees. With their reduced dependence on trees we can imagine *H. ergaster* stepping out into plains largely devoid of trees. Such countryside no longer posed a barrier to their dispersal.

There is considerable uncertainty, and indeed some dispute, about the classification of many of the fossils that have been dated over the million or so years after 1.8 Mya. In part this is due to the fragmentary nature of most of them. A straightforward approach, that I will follow, is that of anthropologist Richard G. Klein of Stanford University. He refers to those found in the African continent, from Morocco in the west, to Ethiopia in the east and down to South Africa, as *H. ergaster*, and only uses *H. erectus* for Asian specimens. A fossil of *H. ergaster*, dating from 1.7 Mya has been found in Georgia, Asia, and the larger brained *H. erectus* probably evolved from this species spreading as far east as China and Java. It represents the first 'out of Africa' migration of *Homo* species; there were to be at least two more. *H. erectus*, as recognized by Richard Klein, first appeared in east Asia about a million years ago and existed until comparatively recently. By and large it seems that over this period there was not much sequential change in the bone structure.

Whereas Acheulean tools have been found frequently in Africa, Europe, and west Asia, none of that age had been found in east Asia until a very few years ago when some were found in China and dated to 800,000 years ago. They are of a somewhat different design to the Acheulean axe of *H. ergaster*. Perhaps their rarity indicates that the Erects made more use of wooden tools that would not have been preserved. Fossils found recently in central Java that may be as late as 27,000 years ago are considered to be those of Erects. They would have been contemporary with modern *H. sapiens*. As these fossils are most unlikely to be the last Erects that lived in Asia, we can speculate that the legendary Sumatran man-ape, the orang pendek, may be based on distant folk memories of the Erects, reinforced by partial sightings of contemporary animals or their distorted footprints.

Sometime in the period 800,000 to 100,000 years ago human evolution made another spurt or series of spurts, giving rise in Africa, and in adjacent Europe and Asia, to what we may term archaic *H. sapiens*. Evidence for such development comes from material found at Atapuerca, Spain, dated around 780,000 years ago[3], which is sufficiently distinct to be regarded by many experts as another species (*H. antecessor*). In addition, in Africa fossils from Broken Hill (Kabwe), Zambia, from around 400,000 years ago, are thought by many to represent the same species as modern humans (*H. sapiens*), though differences have caused them to be put in a loose group 'archaic

[3] Other material from the same area has an age of 325–205 thousand years.

sapiens'. Within the last half million years some of these different hominins evolved into what are now regarded as two species: *H. neanderthalensis* and *H. sapiens*. But before considering them we should pause to reflect upon some aspects of human evolution.

Spurts in human evolution

In 1976 evolutionist Stephen Jay Gould built on an earlier idea to suggest that human evolution was not a ladder, but a jerky process better represented by a bush (FIG. 11.3). As we have seen, hominin evolution seems to have consisted of a series of spurts when a number of major changes occurred. Gould proposed that these changes were due to changes in the rates of development of various parts of the body—termed heterchrony. Humans have retained many of the features of young apes (FIG. 11.5), including a larger ratio of brain to body size and small canine teeth, while becoming sexually mature. This particular phenomenon, sexual maturity while retaining the juvenile form, is known as neoteny and is found in various parts of the animal kingdom; it normally results in larger individuals and longer life-spans. This theory neatly explains many human features and how evolution could occur quickly through changes in the genes that control the development of various parts of the body. Humans are sexy juveniles!

All was not straightforward on this evolutionary path. The increasing size of the brain conflicted with the contrary pressure to narrow the hips in order to facilitate bipedal walking. The birth process in australopithecines would have been easy, as it is in apes; but in humans birth is normally

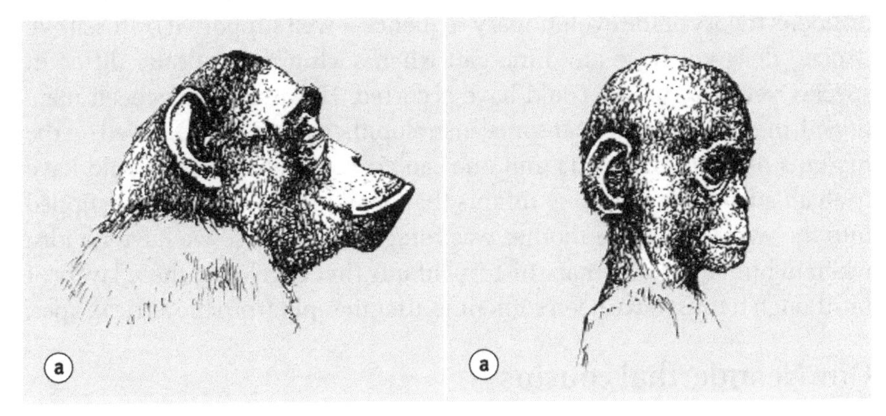

FIG. 11.5 Sketches to show facial proportions of (a) mature and (b) young chimpanzees (from photographs in Naef, 1926)

difficult and involves the rotation of the baby in the birth canal. It seems that the human baby has reached the limit to its size and it has been argued that in relation to the baby's development it should remain in the mother for 21 months. Indeed the human baby is very helpless at birth compared with the young of apes and following this line of argument it has been described as an 'extra-uterine embryo'. In this stage its survival depends on almost continuous care by its mother and, if maternal care is not forthcoming it is liable to scream inconsolably. Such behaviour in the wild could attract predators and, in similar circumstances, young monkeys remain quiet. We can only suppose that, by the time this habit evolved, desertion was a greater threat to a newly born human than were predators.

A contrary argument as to why the human baby is so noisy has been put forward on the basis of an unorthodox theory that humans were at one time aquatic apes. Babies have layers of white fat, far greater in quantity than any other primate. There is no reason why the human baby should need more energy reserves than these others and recent work suggests that this fat is not particularly good for insulation; white fat is, however, effective in giving buoyancy. Adding to the argument, its protagonists, especially writer Elaine Morgan, point out that babies can sometimes swim before they can toddle, that many women find it relatively easy to give birth in water, and that we are largely hairless (like aquatic mammals). Thus, the baby is much more helpless on land than it can be in water; that—it is argued—is why parents have to carry babies and why the baby yells if parted from its parent. It is an intriguing idea but there are many facts against it, for instance, that hominoid fossils have not been found with other marine organisms and that the orthodox theory of the evolutionary sequence—well supported by fossil evidence—does not have any time gap when such a dramatically different species, yet to be found, could have occurred. However, as has been mentioned, there is evidence that some australopithecines may have lived on the margins of lakes and rivers and one can speculate that there would have been an advantage in having infants that would float easily if they slipped into the water when the mother was foraging. Of course we have no idea when in human evolutionary history infants first developed these layers of fat, though it must have been sometime after the split from the African apes.

Our Neanderthal cousins

Much hominin fossil material found in Europe and west Asia that comes from the periods between about 300,000 years ago and 30,000 years ago differs from that recognized as being modern man in a number of features. These include greater sturdiness of limbs, more prominent brow ridges, the larger nose and brain case and the smaller chin. Associated with them are many varieties of hand tool, that are more elaborate than the Acheulean, numerous small flakes having been chipped off around the edge (FIG. 11.4); the principal form is known as Mousterian. These hominins are known as Neanderthals, as the first remains were discovered in the Neander valley, Germany; today they are generally designated as a species, *H. neanderthalensis*. The most northerly Neanderthal remains found are from North Wales (Pontnewydd), the most easterly from Uzbekistan (Teshik-Tash) and the most southerly near Gibraltar and in southern Israel—this whole area is sometimes referred to as 'Neanderland'. During the Ice Ages this land would have had a cold and mostly dry climate, much of it coniferous forest. To the north lay the glacial or tundra steppes, a habitat that does not exist today, but would have had grazing mammoths, straight-tusked elephants, woolly rhinoceros, horses, reindeer, and bison. Deer would have browsed in the forests and goats and sheep on mountain slopes. All of these were potential food.

Neanderthals certainly lived in caves where these were available, but they may well have moved camp with the seasons, while retaining a home base. Judging by the rocks used for the tool-making, each group had a relatively small territory, perhaps some 7000 hectares. Whether their weapons were sufficiently sophisticated for them to kill large mammals is doubtful, but collections of bones show that they ate them. Very often the larger bones have been cracked open to remove the marrow. In some situations it seems they used natural features such as cliffs or bogs as traps, driving animals into these so that, if the mammals were still alive, they could be more easily killed. It has also been suggested that they may have scavenged the prey of predators such as cave lions, hyenas, or wolves, perhaps even driving them off their kill.

Their life was certainly risky and tough, for very many of the bones show signs of old injuries that had healed. One shows evidence of a stab wound caused, it would appear, by the frontal attack by a right-handed assailant, though the victim seems to have lived for some weeks, presumably cared for by other members of the tribe. As whole skeletons have been found, often lying on their sides with their legs drawn up, they must have buried their dead and protected the graves from scavengers like hyenas. Without deliberate burial wild animals or vultures would feast on the corpse and the parts of the skeleton would have become scattered. It is difficult to interpret the significance of plant and animal remains and the unusual stones sometimes found in graves, but some seem to have been placed deliberately. These may signal that they had what we would term religious beliefs. There is also some evidence from Neanderthal bones which show cut marks or signs of butchering, that Neanderthals were cannibals, but the circumstances in which this occurred are of course unknown. Was it a time of stress? Had the victims already died? Were they strangers or members of the group?

Neanderthals could talk, but their range of sounds may have been more limited than in modern humans. This view is based on the detailed anatomy of the skull that indicates that the larynx was high in the throat as it is in a baby. The inside of the skull reflects the structure of the brain and shows that the areas associated with speech and comprehension were more restricted than in *H. sapiens.* On the other hand one skeleton has been found with a hyoid (tongue) bone intact; it was identical to that of modern humans, where it plays a key role in voice production.

As pointed out already *Homo* is virtually unique among primates in living in cold climates and, although natives of Patagonia were noted by Charles Darwin as living naked in glacial conditions, it seems likely that Neanderthals would have wrapped themselves in the skins of animals. There is some evidence for this. Stone scrapers are common and the wear pattern often found on the front teeth suggests they were used as a vice to hold a skin while the fat and other material was scrapped off. Hearths are common in their caves and the fires would have served for cooking, heating and deterring dangerous animals. Fire-making would have been an important technology for the Neanderthals, allowing them to colonize colder regions after they or their ancestors (probably *H. ergaster*) formed the second wave of *Homo* 'out of Africa'. Nevertheless, unlike *H. sapiens,* they were not able to colonize the coldest regions of central Eurasia.

Around 100,000 yeas ago Neanderthals living in the Levant (the coastal regions of Turkey-in-Asia, Syria, Lebanon, and Israel) encountered humans with different features, namely those of *Homo sapiens,* that had drifted up from Africa. These appear to have been similar to the fossils found at Broken Hill, Africa—like them they are grouped as archaic *H. sapiens,* but often termed Cro-Magnons. The information comes from material found in caves in Israel; that dated around 100,000 years ago from Quafez and Skhul is of *H. sapiens,* whilst that of a Neanderthal from Tabun is about the same age. The Neanderthals lived on (or returned), for their remains in Amud and Kabara date from around 50,000 years ago. Thus for 40–50,000 years the two coexisted in the Levant until, as will be described, in another and final 'Out of Africa' invasion of Europe by *Homo* species, the Cro-Magnons moved up through the Levant and into Europe. By about 27,000 years ago the Neanderthals had become extinct.

Ourselves

There are two theories as to the origins of modern humans. One is that the earlier hominins in various parts of Africa, Eurasia, and perhaps Australia, evolved more or less 'in step' to give the human population today, this is the multiregional model. The other is that we all arose from some subpopulation of hominins around 200,000–100,000 years ago, probably in East Africa and hence this theory is often termed the 'Out of Africa' theory. In its most straightforward form this latter theory supposes that there was no hybridisation—no gene flow—between the archaic *H. sapiens* and the other hominins that they would have encountered as they expanded out of Africa. There is, however, some evidence that could be interpreted as indicating that this did occur to a very limited extent. If so, the stark contrast between these two theories is less—neither is completely wrong nor totally right. I will outline this evidence later (p. 231).

The remains of modern humans can be recognized by their anatomical features, such as the high, short brain case, the vertical forehead, the brow ridge reduced especially in the middle, and the more prominent chin, and are dated around 100,000 years ago. However, at least another 50,000 years passed before the tools and other artefacts associated with modern man appeared. This was a very key time, the time when biological evolution virtually stopped and cultural evolution took over—it has been termed 'the great leap forward'. It seems likely that the evolution of complex language, the transmission of knowledge and the greater development of the ability to guess what another is thinking (theory of mind) were the facilitators of this development.

The distinctiveness of the 'tool-kit' of early modern humans has long been recognized by archaeologists and it has been termed the Upper Palaeolithic. The stone tools are elaborately worked, often much longer than wide and with a sharp point (FIG. 11.4), while others served as borers or scrapers. However, two further features distinguish them from the preceding Mousterian (Middle Palaeolithic) that is associated with the Neanderthals. Firstly, materials other than stone were often used, especially bone and antler. Secondly, there were objects that had no practical use, and which were clearly used for ornamentation. Art had emerged, and by 32,000 years ago figurines were carved and the walls of caves decorated. Many of the tools were clearly hafted, attached to handles often of wood. Very recent comparisons of the structure and wear of the hand bones of Neanderthals and archaic *H. sapiens* from the Levant indicate that, although as yet no striking difference has been found in their tools, the hands of Neanderthals were adapted for action by the direct transfer of power such as when holding a hammerstone or hand axe without a shaft; while those of the archaics would function best when obliquely transferring power, as when holding a hammer by its handle. This variation in grip may have opened the way for the development by *H. sapiens* of the varied types of tool found in the Upper Palaeolithic.

In 1994 research using the approaches of modern molecular biology started to add to this picture. Becky Cann and her colleagues from the University of California studied the composition of the mitochondrial (non-nuclear) DNA—inherited only through the female line—from a wide range of modern humans. They concluded that we had all evolved from a single female ancestor about 200,000 years ago—hence this ancestor is referred to as the 'Mitochondrial Eve'. They believed that the mathematics of their analysis showed that this ancestor had come from Africa. Other studies, including some based on the Y-chromosome, which is inherited through the male line—confirmed the comparatively recent common origin of humans living today. The genetic variation in *H. sapiens* is exceedingly small, less than that in one subspecies of chimpanzee and it has been calculated that at one time the population of those humans who are our ancestors numbered around 10,000 individuals. Around 50,000 years ago there was a rapid increase in population. This is only about 2500 generations. An indication of the shortness of this in biological terms is shown by the comparison with house mice, a colony of which will have had the same number of generations since the 1690s when William and Mary were King and Queen of England and a college named after them was founded in Virginia. It is therefore little wonder that in fundamental genetic make-up we are all so similar.

Whereas further analyses of the data from the early mitochondrial studies suggested that the origin could not be proven as being in Africa, studies of the complete mitochondrial genome, published at the end of 2000, showed this to be the case. Thus, these genetic studies strongly favour the 'Out of Africa' theory, but the matter is still highly controversial.

It seems possible that the ancestors of modern humans left Africa on more than one occasion. Perhaps an early migration was across the southern end of the Red Sea when sea level would have been lowered because of the water held in the ice caps during the first part of the Last Glaciation. Alternatively those in the Levant may have moved out into Asia. These migrations, starting around 80,000 years ago, would have continued across Asia to China. Some eventually made the sea crossing to Australia and their descendants are represented by 'Mungo Man', recently (but controversially) dated as 62,000 years old. Others, as we have seen, remained in the Levant (which was effectively an extension of Africa) then moved north and across Europe some 40,000 years ago, sweeping back into Asia and eventually into the Americas some 13,000–12,000 years ago.

This is highly speculative and we can expect many new findings from research using the techniques of molecular biology and from improved dating of archaeological material, as well as discoveries of new fossils, to modify the story. If one accepts the 'Out of Africa' theory a most controversial question is this: 'Did modern humans have sexual intercourse with any of the hominins they encountered on their migrations?' The differences between *H. sapiens* and the Neanderthals was no greater than that between two subspecies of chimpanzees, and the distinction with the Erects was little more. On a purely biological scale they were all very close and the production of fertile hybrids was very likely. However, there may have been cultural or behavioural barriers.

Although certain remains have been found that have been claimed to be intermediate, their interpretation is uncertain. What is rather less controversial is that about 40,000 years ago, when modern humans and Neanderthals coexisted, the collections of tools and other objects associated with the Neanderthals resembled those of the contemporary *H. sapiens*; for example, there are objects that appear to have been used for personal adornment. These Châtelperronian artefacts (from central and SW France and NE Spain) suggest that there was some cultural exchange between the two groups. Historical accounts of the behaviour of explorers and invaders show that, unless there were great cultural barriers, some genetic mixing, whether by rape or courtship, was likely. However, most evidence from molecular biology suggests that if this occurred it was on a very small scale; we can be confident that the Neanderthals did not contribute to the mitochondrial genome. But some research workers have identified certain nuclear genes that appear to represent a very ancient variation and so could be derived from another gene pool. The gene for red hair seems to be one such and as it was originally found only in Europe there have been speculations that it came from the Neanderthals. Another slightly anomalous gene is believed by some to be found in the Andaman Islanders, while mitochondrial DNA recovered from Mungo Man is believed to have an extinct, but archaic component. Research in the next decades will certainly change our understanding of this, and many other, aspects of human evolution.

It is undoubtedly our ability to modify our environment that enabled modern *H. sapiens* to spread round the world and successfully occupy the lands of the Neanderthals and the Erects, as well as those they had not colonized.

CHAPTER 12

Humans: The Great Modifiers

40,000 years ago, Today and Tomorrow

ALL organisms have an impact on their environment. Some, ranging from coral polyps forming reefs to beavers forming dams, have a very real effect on the physical environment. But the impact of *Homo sapiens* since the 'Great Leap Forward', 50–40,000 years ago, has been more varied and has become increasingly extensive; it is, however, the key to our success—our ability to exist in vast numbers and to live in virtually every corner of the globe. The Neanderthals existed for over 300,000 years, many times longer than modern man has yet achieved, but they never extended beyond southern Europe and South-west Asia (Neanderland); their range was limited, even when allowing for the ice sheets.

Technological innovations have allowed humans to alter the ecology of the environment in ways that have benefited them. This is well illustrated by the changes in what is termed the carrying capacity. This is an ecological measure of the number of a particular organism that can be sustained in a unit of habitat; if the carrying capacity is exceeded, famine normally results and sometimes the habitat can also be damaged so that the carrying capacity is permanently reduced. The story of Easter Island is an example of human overexploitation (p. 245). Assessments of the carrying capacity achieved by various 'farming' styles in terms of the number of persons that can be fed from one square mile of land show a dramatic increase with advancing technology:

hunter–gatherer (Tropical rain forest) 1;

shifting cultivation 15;

medieval agriculture 50;

intense tropical peasant agriculture (New Guinea) 125;

modern intensive agriculture (Western Europe) 400.

This great increase in carrying capacity has been brought about by three features: the increasing sophistication of tools, from arrows to aeroplanes; the domestication of certain plants and animals; and the extensive use of minerals (fossil fuels, metals, clay, etc.) directly or to make new materials, including artificial fertilizers. Just as corals, trees, and termites create ecospace for themselves and others, so do humans with their buildings, whether huts or skyscrapers. This technological progression is represented by a succession of lifestyles: hunter–gatherer, agriculturalist and, finally, urban-dweller and industrialist. Obviously the transformation of the earlier lifestyles into the later ones has been gradual and occurred at different times in different parts of the world. Hunter–gatherer societies still exist and agriculture (food production) remains the essential source of food even when most of a population follows an urban lifestyle.

The hunter–gatherer

There is evidence that the Neanderthals ate a lot of meat, though the extent to which they actually hunted big game is uncertain, for unless their stone axes were attached to handles this would have involved very dangerous close combat. Their remains do, however, show many signs of bones that had been damaged and healed, and this contrasts with the relative freedom from fracture in the Cro-Magnons. Undoubtedly the dangers in hunting large mobile animals were much reduced when weapons that maintained a distance from the animal—spears and bows and arrows—were developed by the Cro-Magnons. This probably happened by at least 20,000 years ago and within 6000 years a variety of projectiles as well as fish hooks were in

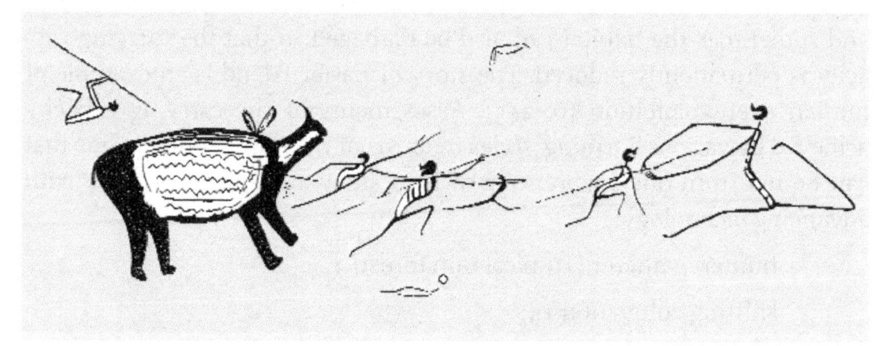

FIG. 12.1 Mesolithic (12,000–10,000 years ago) rock painting from Bhopal, India, showing an apparently unsuccessful hunt of a rhinoceros, although the larger figure on the right may represent the hero who reversed the fortunes (from Neumayer,1983)

use. Dogs, domesticated around 12,000 years ago, would have further increased the ability to hunt 'at a distance' and lessen the dangers. The importance of hunting and the types of animal hunted is shown by the representations in cave paintings in southern Europe, some of which (such as the Cassac Cave in the Dordogne) are believed to date from as early as 35,000–30,000 years ago, and in rock paintings elsewhere (FIG. 12.1).

During the same period that humans were perfecting their hunting techniques many large animals became extinct (FIG. 12.2) and it seems very likely that this was cause and effect. The Bering land bridge, between Siberia and Alaska, was crossed by humans around 12,000 years ago and within the following thousand years the mammoths, other species of elephant, a giant bison, four species of camel, and another 28 genera of large mammals were extinct. Also within a thousand years, humans had colonized South America and there too this coincided with the extinction of the giant ground sloth over five metres long, an armadillo nearly 3.5 metres long and weighing two tons, and some 44 other genera. Against that 'blitzkrieg' theory it has been argued that this was the very coldest part of the Last Ice Age followed by a warming (12,000 years ago), a further cooling (the 'Younger Dryas Stage', 11–10,000 years ago) and a final warming. These climatic conditions could have stressed the populations of many animals and could have been the cause of their extinction. However, they had survived the climatic fluctuations of earlier Ice Ages and there does not seem to have been a general extinction. The animals affected were just the very large ones (which would have had slow rates of reproduction) and this extinction occurred particularly on those continents where *H. sapiens* had just arrived.

The picture is rather less clear in Australia. Here too there was a fauna with giants—a two-ton wombat, a 2.5 metre high kangaroo and a giant bird, *Dromornis* standing three metres high and weighing half a ton. Altogether 15 out of 16 genera of large animals became extinct, probably around 14–13,000 years ago. As we saw in the last chapter there is some evidence that *H. sapiens* lived, around Lake Mungo at least, at a much earlier period. There are many uncertainties surrounding this possible coexistance. Is the new date for Mungo Man correct? If so, were humans very restricted in their distribution in Australia until later, possibly following a second invasion? Did Mungo Man and his compatriots have the hunting techniques to tackle giant animals safely or were these brought by a later wave of invaders? When did man-made fire become a major influence in the ecology of Australia ?

FIG. 12.2 Large animals of Australia (a)–(c) and South America (d)–(g) that were probably driven to extinction by over-hunting in the Pleistocene:
(a) Giant kangaroo *Macropus ferragus*; (b) Giant short-faced kangaroo, *Procoptodon goliah*; (c) Giant wombat, *Diprotodon optatum*; (d) Giant armadillo, *Glyptodon*; (e) Litoptern, *Macrauchenia*; (f) *Toxodon*; (g) Giant ground sloth, *Megatherium* (modified from Stuart, 1986 and (d) from Savage and Long, 1986)

Extinctions were less extensive and less simultaneous in Africa and Eurasia, which lost two and seven genera respectively. It has been pointed out that in these continents humans and animals evolved together and so it is suggested that as humans became more dangerous so animals became more wary. In the other continents 'fully-armed' men accompanied by dogs encountered animals that were naïve. These animals did not recognize the danger posed by mammals so much smaller than themselves.

There is also more recent evidence that supports the interpretation that these Pleistocene extinctions were due to overkill. Humans reached Madagascar and New Zealand around 1000 years ago; the extinction of giant lemurs and elephant birds in the former, and of moas (FIG. 12.3) in the latter, soon followed. Within historical times there are even more precisely dated episodes. Mauritius was colonized in 1638 and the last Dodo (FIG. 12.4) was killed in 1681. The seven-metre-long Steller's sea cow (a relative of the manatee, p. 209) was discovered on the islands in the Bering Sea in 1742 and was extinct by 1769. Contemporary accounts show that these large animals were not alarmed by the approach of humans—they were naïve. If a Steller's sea cow was wounded others would crowd round apparently seeking to provide protection, but with man as the predator they too would have fallen victim.

Low rates of reproduction are a feature of large birds and mammals; they often take many years to reach maturity and then produce few young. For example, the Californian condor, with a wing span approaching three metres, is eight years old before it breeds and then a pair produces only one egg every other year. These animals are, however, long-lived (up to 45 years in the case of the condor), but the lengthy lifespan is necessary for the population to replace itself. If circumstances change, such that the life expectancy is shortened, these large animals are at risk of extinction. Hunting was probably the major cause of these Pleistocene extinctions, but the change in climate

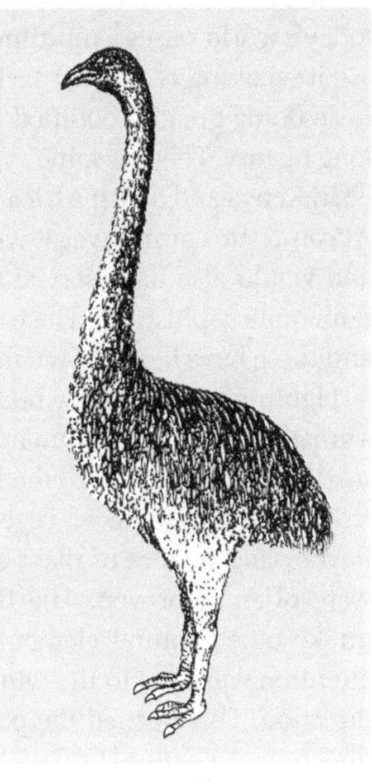

FIG. 12.3 A moa, *Dinornis*

could have increased the animals's vulnerability. Both factors combined would have lowered life expectancy.

Bows and arrows, and spears remained the main hunting tools for well over 12,000 years, until in the last few hundred years fire-arms were invented. As these have spread round the world, so has the risk of extinction for other life. The harpoon gun exposed the giants of the sea, the whales, to unsustainable levels of predation (now hopefully put to an end by international treaty). On land, extinction risks rise when 'traditional hunters' lay aside their bows, spears, and blow pipes and take up fire-arms. Deaths due to hunting are likely to reach a level that the prey population cannot sustain. The continuation of the hunter-gatherer lifestyle in today's world raises a multitude of often contradictory ethical questions.

FIG. 12.4 The dodo, *Raphus cucullatus* (after Herbert, 1634)

Fire-making is another technology that hunter-gatherers have used and in so doing greatly modified their environment. Like stone axes, this has a long history. There is some evidence that 1.4 Mya fire was being manipulated in Kenya and South Africa and it was certainly used by the Neanderthals. At camp-sites or in caves it would have been used for heating and cooking, and would also have served to ward off large animals. The Cro-Magnons built quite sophisticated hearths. These had small ditches leading into them and these have been shown to function by increasing air flow.

Lightning would cause bush fires and, although these must have proved a great danger to early humans, we can envisage that four observations were made: fire travelled with the wind and needed plants to burn, animals fled from it; a new type of vegetation eventually grew after a fire; and the flavour and texture of plant and animals that had been cooked in the fire were often improved. The first would have led to the recognition that, unlike other natural elements (wind and rain), fire could be 'herded': if attention was paid to the wind the fire could be made to go in a particular direction. This opened the possibility of using fire to drive animals in the direction of a natural trap such as a cliff or bog. It could be an important part of a hunting strategy.

Fires have a profound effect on vegetation; regular fires can maintain a particular type of plant community that would be replaced by another in the absence of fire; such a community is termed a fire climax. Plants of fire climaxes have evolved adaptations such as the ability to regenerate from parts protected below ground, while some trees have fire resistant bark. There are other species whose seeds will germinate only after being heated in a fire, while others contain inflammable oils so that the bush or tree virtually explodes when the fire approaches. These adaptations give them an evolutionary advantage as the fire destroys seeds and seedlings of other bushes and trees which, if left, would eventually out-compete them.

There is much debate as to whether various fire climaxes that exist today arose millennia ago simply due a different climatic regime or whether fire was the key to their origin and, if so, was this natural lightning or man-made fire? The effect of the fire depends on its temperature and this relates directly to the amount of combustible material in the fire's path. Fires quickly pass through grassland and certain types of woodland, leaving the trees unscathed, but destroying any young bushes or saplings. There follows a flush of grass and herbs—thriving on the nutrients released by the fire—that is most attractive to grazers and to the hunters, for there are no bushes to obscure the view. On a much larger scale are the savannahs of Africa, the prairies of North America and the pampas of South America, all of which seem to be maintained by frequent fires. It is likely that for many millennia most of these have been started by humans. The types of vegetation that resulted would attract herbivores for hunting, while the open vista would reduce the risk of ambush by large predators or enemies.

Where there is little summer rain a shrubby type of fire climax develops, such as the maquis of the Mediterranean, the sage bush of western USA, the 'bush' of Australia and the fynbos of South Africa; here too fire will aid the hunter: driving animals, reducing cover and stimulating new plant growth.

Little is known about the plant material that was gathered by early *H. sapiens* but, as well as fruits, bulbs, and tubers, the seeds of grasses and pulses were likely to have been collected for they were to play a major role in the next lifestyle.

The agriculturalist

Hunter-gatherers living in seasonal climates must have frequently experienced periods of food shortage, when game was rare and fruits and seeds hard to find. These difficulties would be ameliorated if edible plants and animals could be concentrated and stored near the home base, particularly if they were there during the period of shortage. The development of agriculture had these benefits. Essentially there were two stages: firstly, working with wild species and secondly, the selection of particularly suitable forms or varieties—domestication. Agriculture would have started as a supplement before it became a way of life. Thus, the development of agriculture would have been an evolution not a revolution; it is associated with the Neolithic period in which polished or ground stone tools appeared.

The practice of agriculture developed, almost certainly independently, in two, if not four, regions. In Eurasia the earliest cultivated wheat was grown in the Near East (Turkey through to Iraq) around 10,000 years ago, but there was probably another independent development in the Far East (north-east China and Japan) where rice, millet and gourds were brought into cultivation some 2000 years later. It seems likely that there were also two independent developments in the Americas. Squash and bottle gourd were grown in Oaxaca, Mexico 10,000 years ago; maize was grown near there from 7000 years ago. By 3500 years ago both these had reached south-western USA—maize seems to have travelled faster than gourds. Beans were not cultivated until later—2300 years ago in Mexico. In South America, in the Andes and western Amazonia (Peru and Ecuador), maize was cultivated 6000 years ago; the chilli bean, the potato, and the pepper 2000 or more years earlier. Over 2–3000 years later additional crops and other varieties were brought into cultivation; in Eurasia, for example, einkorn wheat, barley, peas, and lentils.

Animals were also domesticated. Almost all domesticated animals are pack or herd animals and one can envisage that domestication occurred when a young animal, whose mother had been hunted and killed, was kept alive. We do not have to attribute any softness of heart to the captors—at this stage the intention was probably simply to allow the young to grow further to make a bigger meal. However, being a herd animal it would come to regard its captors as its herd and become tame. This story is told in the well known nursery rhyme:

Mary had a little lamb
Its fleece was white as snow
And everywhere that Mary went
The lamb was sure to go.

Domesticated animals have more fat, are smaller (FIG. 12.5), lighter in colour and less aggressive than the wild species from which they originated.

FIG. 12.5 The wild auroch bull (above) compared with its domestic descendant (after Clutton-Brock, 1981)

In part, this was probably due to selection: lighter animals could be more easily seen, especially at dusk or during the night, when they needed to be guarded—a point referred to by the Roman writer Columella; less aggressive animals were clearly desirable, and smaller ones could be more easily controlled. It is also possible that smaller individuals survived better on the inadequate rations provided by the early farmers. Cattle corralled at night to prevent wandering and to protect them from predators—including two-legged ones—would have had their grazing time restricted.

Domesticated animals fulfil a number of functions in human societies: they may be a source of food, of materials (skins, wool, horn, grease), and of muscle power (to move things), as well as a biological weapon in hunting; examples of the latter are dogs, horses and falcons. Additionally in many societies they may have a social role—in religion, display or sport or simply as pets.

The dog was the first animal to be domesticated—some 12,000 years ago, ahead of the advent of agriculture. As has been mentioned already it provided, and still provides, an important weapon for the hunter allowing him to harry the prey while remaining at a distance. The origin of the dog is undoubtedly the wolf, probably the Asiatic Wolf that is smaller than most other species. In the dingo, taken to Australia by man 10,000 or more years ago, we probably see an animal close in appearance to the original domesticated dog. Dogs have a much greater tendency than wolves to bark and this is essential for one of their major roles—as watch dogs. This function has been characteristic of dogs through the ages so that certain organizations that have a role in guarding society against particular risks are described as 'watchdogs'. When humans crossed the Bering land-bridge into North America around 12,000 years ago, the dog was their only domestic animal and, unsurpisingly, it seems they took it with them. In the north the dog was often used as a beast of burden, but in Mexico as a source of meat; the edible dog was a particularly bulky animal.

The major ruminants (goats, sheep, water buffalo and cattle) and the pigs, were probably first domesticated around 10,000–8000 years ago. Molecular studies of the mitochondrial DNA of present-day breeds show that in all those listed, other than goats, there were two separate domestication events, one probably in south-west Asia and the other in far eastern Asia. Goats appear to have been domesticated on three occasions: once around 10,000 years ago, the descendants of this event are very widespread; once around 6000 years ago whose line is found in certain rare breeds in Mongolia; and most recently 2000 years ago and now represented in west Pakistan.

The domestication of the horse started around 6000 years ago—but it seems that it happened on many occasions, with wild mares being captured and broken-in. Whereas with the ruminants and the pig we can envisage animals being spread by trade, with the horse it was often the technique, of capturing and taming the horse that spread through Eurasia. The horse provided humans with a means of travelling at speed over great distances; its domestication was the first technological step towards 'shrinking the world' and its role in warfare has had major political repercussions down the centuries.

A wide range of other animals have been used by humans as beasts of burden, hunters or sources of meat or materials. These include elephants, reindeer, camels and, in South America (where all these other species were absent) llamas and alpacas. Generally, these domestic animals are not very different from the wild form. Indeed often they are recruited from wild herds and distinct domestic breeds do not exist.

The one significant domestic animal that is not a herd animal is the cat but then, as cat owners will testify, the cat retains much of its independence and simply deigns to live with one! For most of the history of domestication (and in many circumstances today) other animals, as well as cats and dogs, lived in close proximity with humans and many diseases were transferred, for example, from cattle humans developed tuberculosis and cowpox—evolving to small pox—and influenza probably came from pigs. When the Europeans, who by then had developed some resistance to these diseases, travelled to other parts of the world—the Americas, Australasia and many islands, where domestic animals and their diseases were absent—these infections with animal origins took a heavy toll of the local inhabitants. The political dominance of the European invaders owed as much to diseases they brought with them as to their weaponry and martial skills.

The cultivation of crops and the grazing of domestic animals, in other words, farming, alters the ecological system (ecosystem). In the first place, the plants have often been selected by the farmer, so are not those that are best adapted to flourish under the local conditions. In planting the crop, the soil is disturbed and there is often much bare soil. This soil is then vulnerable to erosion by wind or water. Until cultivation methods were altered in the Corn Belt of the USA as much as 70 cm of soil could be lost in a single year. The vegetation of a crop is altogether sparser; in wild grasslands in England there are commonly 100 million grass stems per hectare, but the cereal crop that replaces it will have no more than 4 million stems in the same area. In traditional farming, such as the gardens in Papua New Guinea, several different crops will be grown together and this has now been shown to reduce the impact of pests. But in modern agriculture only one type of plant is grown—it is a monoculture: the plants are all the same age and of the same genetic make-up which makes them very susceptible to outbreaks of pests and freak weather conditions. Natural vegetation normally consists of many different types of plant and even those that belong to the same species will often be of different genetic make-up. This diversity reduces the risk of a pest outbreak. In crops, species other than the crop plant are eliminated and modern herbicides do this most effectively. These weeds, however, provide the food for a variety of insects, so herbicide applications in cereal fields have reduced the number of insects, perhaps to one-fifth of their previous level, and this has affected some farmland birds (e.g., the partridge) that depend on those insects for food. In fact, the use of various pesticides in modern agriculture has had a profound influence on the other organisms, from soil fungi to birds.

Farmers will often supplement the supply of water and minerals in the soil, both of which can cause environmental problems. Irrigation can lead, in many climates, to salt being left in the soil as the water evaporates;

eventually ordinary crops can be grown no longer. Fertilizers are often washed out and the excess nitrogen and phosphorus cause massive algal growth in adjacent waterways, leading to anoxic conditions and the death of fish—a phenomenon known as eutrophication.

Though we may point to reductions in biodiversity as being due to the intensification of agriculture over the last half century, a lot of the diversity that is being lost is actually due, in the first place, to the practice of agriculture. Prior to human activity much of the world was covered by forests (wild woods). Initially, some of these were cleared by hunter-gathers using fire and stone axes and the practice was undoubtedly extended by the early farmers. We can tell that these wild woods, like woodlands today, had a limited number of different plants from a study of pollen in ancient deposits. If the activity of humans and their domestic animals ceases, such woodlands reappear. One well-known demonstration of this is at Rothamsted Experimental Station, Harpenden, UK, where a piece of a wheat field was fenced off and left alone in 1882—this 'Broadbaulk Wilderness' is now an oak wood. In many temperate regions, particularly western Europe, the activities of farmers led to the creation of a countryside which was a patchwork of small fields and uncultivated land, including woodlands. Such a mixture of habitats will have a diverse flora and associated fauna. A whole range of species that would have been very rare in a forested landscape could flourish in this countryside. The skylark is an example. The change in the last century is due, not only to the widespread use of pesticides, but to the advent of the chain saw, the tractor and the bulldozer that enable natural obstacles (wood lots, hedgerows, large rocks) to be easily cleared and the sizes of fields increased—the patchwork is greatly simplified.

The process of clearing natural woodland is still occurring in tropical rainforests where there is a natural type of patchwork due to the enormous variety of trees (p. 180). Using fire and modern equipment, large areas are being totally cleared. A large number of species of animal and plant are becoming extinct—many even before they have been discovered.

Returning to prehistory, farming allowed more people to live in an area and the provision of a more reliable food supply would have led to greater survival and population growth, though some individuals were of poorer physique. As this process continued more trees were removed for fuel and other uses and to make room for crops. More animals were kept and the vegetation grazed and browsed more intensely, sometimes so much so that it was killed. In other words the environment was overexploited. The community would then collapse. On a continent the people and their animals could move to pastures new—'a land flowing with milk and honey'.[1] There is evidence that many regions of north Africa and western Asia, now largely deserts, were formerly richly covered in vegetation. However, there is controversy as to whether this was entirely due to human overexploitation or at least in part accelerated by a natural change in climate.

FIG. 12.6 Carved head on Easter Island

The human influence can be more clearly established in Easter Island, where numerous carved heads (FIG. 12.6) show that there was previously a thriving community, and analysis of the pollen in the soil shows that there was luxuriant vegetation. The history seems to have been that the island was colonised by the Polynesians about 1600 years ago (600 AD), a time when it was covered with palms and other trees. Within a thousand years the human population had grown to about 7000, the forests had been cleared and the land intensively cultivated. This civilization prospered for about 280 years, during which period the many sculptures were made. The intensity of cultivation resulted in soil erosion, smaller crops and a lack of large tree trunks to make sea-going canoes. The latter also contributed to the food crisis as, apart from chickens, fishing was the only source of protein. Whereas on continents members of a population that had overexploited their environment could move on and invade other lands, without boats the Easter Islanders were imprisoned on the overexploited island; warfare, slavery and probably cannibalism developed, with the result that the population crashed and by 1722, when Europeans reached the island, it was a barren, treeless wilderness; many of the statues had been toppled or never raised, and the stone platforms were derelict. This outcome of population growth and non-sustainable exploitation holds a lesson for the world today.

[1] Exodus chapter 3, verse 8

Urban and industrial people

As agriculture progressed individuals could produce food that was surplus to their families' requirements, and so communities would have formed where some members undertook other tasks, the products of which they exchanged for food—the beginning of urban society and the birth of civilization. This happened in different places at different times. Probably the oldest development was in the Far East, but less is known about this than the Sumerian civilization which flourished in Mesopotamia (modern Iraq) some 5400 years ago. Other civilisations arose later and apparently independently in both North and South America and probably in parts of sub-Saharan Africa, which may have been due to Egyptian influence.

Through the birth of towns, humans brought further changes to the environment; most obviously these were areas where wild animals would not be tolerated, where the vegetation was much modified, and where the waste products of human life and activity accumulated—the beginning of significant pollution. Associated with the development of towns was the further development of industries; initially those involved with metal smelting and working. Copper seems to have been first smelted in Turkey around 6000 years ago and artefacts (pins) made from that metal have been found dating from 4000 years ago, after which the advantage of adding zinc to make the alloy bronze was discovered. Iron started to replace bronze as the most widely used metal from around 3000 years ago, while other metals such as lead, silver, and gold were also extracted. A major step was taken with the advent of the steam engine and the birth of the Industrial Revolution 250 years ago, after which towns grew rapidly and pollution became more widespread. In the last 50 years the scale of industrialization worldwide has increased exponentially, as has the human population (FIG. 12.7). We can categorize the modifications to the environment due to urbanization and industrial activities under three headings: effects on the natural history of towns, restructuring the physical environment, and pollution.

The climate of large cities is slightly different to that of the surrounding area: New York City is at times 3°C warmer than the surrounding countryside. Where towns and cities are largely brick and concrete the fauna and flora will be greatly impoverished, though there are some plants that grow well on old walls and birds, like swifts, starlings and pigeons, nest in or on buildings. Where there are gardens these provide a wide variety of habitats and it has been found in England that the density of nesting birds in

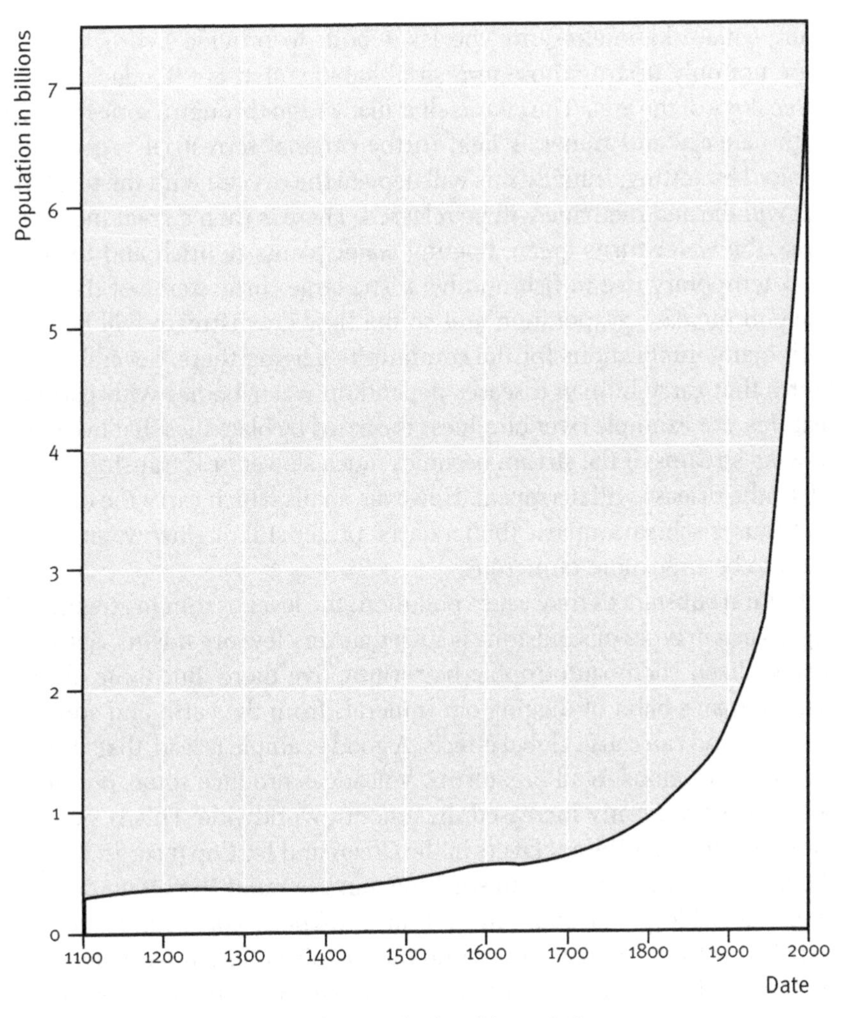

FIG. 12.7 The growth of world population

gardens is greater than that in woodlands and most other 'wild' places. The natural history of towns is different, but overall probably more diverse than that of the adjacent countryside. They normally lack really large animals, but many medium-sized species like foxes and raccoons have adapted to suburban areas.

Humans have long had an impact on the physical form of their environment: pyramids erected, wells or pits dug out, lakes and ponds built. Nowadays, with modern machinery, many such undertakings are on a greater scale than ever before. Particularly spectacular, often on a scale of many square kilometres, are the lakes built to provide hydroelectricity; these not only destroy those terrestrial habitats that are flooded, but alter the ecology of the area. The lake itself is likely to go through a series of stages as the energy and minerals held in the original terrestrial vegetation is released by rotting. Initially this will deplete the oxygen with the result that fish will die and methane will be released. There is then a great increase in algae, the water turns green, floating water plants flourish and there is a final, temporary rise in fish numbers. The large surface area of the lake is likely to increase evaporation and so the flow downstream will be less—with many implications for the communities living there. Several invertebrates that carry human diseases depend on water bodies with particular features. For example river blindness is carried by blackflies that live in fast-flowing streams; if the stream becomes much slower or is transformed into a lake the disease will disappear. However, snails which carry the organism that causes schistosomiasis (bilharziasis) prefer still or slow water and are likely to become more numerous.

Natural substances may cause pollution: the level of iron in streams arising in certain types of sandstone is such that very few organisms, apart from a specialized chemoautotrophic bacterium, live there. But these are local effects; man's habit of digging out minerals from the earth and spreading them around can cause global effects. A good example is lead, that at certain levels is poisonous to all organisms. Volcanoes produce some, but human activities have greatly increased the amount worldwide. This is shown by the quantities in different layers in the Greenland Ice Cap (FIG. 12.8), where the initial increase was due to smelting, but the rapid rise after 1950 came from the widespread addition of lead anti-knock compounds to petrol. Expelled with the exhaust gases into the air this pollutant has been carried by the winds to all parts of the globe. Unlike some chemicals, lead is not quickly washed out of the soil; it has what is termed a long residence time. We breathe it in with the air and get it from our food and drink. Although one has to take in a lot to suffer lead poisoning there is good evidence that lower doses affect behaviour and learning abilities[2], especially in children. Governments in many parts of the world have taken steps to phase out lead additives from petrol. Unfortunately this is not the position in several Third World countries. Lead was formerly used in paint and plumbing and these provide other sources of contamination.

[2] It has been suggested that lead poisoning from drinking wine that was heated in lead vessels played a role in the fall of the Roman Empire, the members of leading families having had their mental abilities impaired!

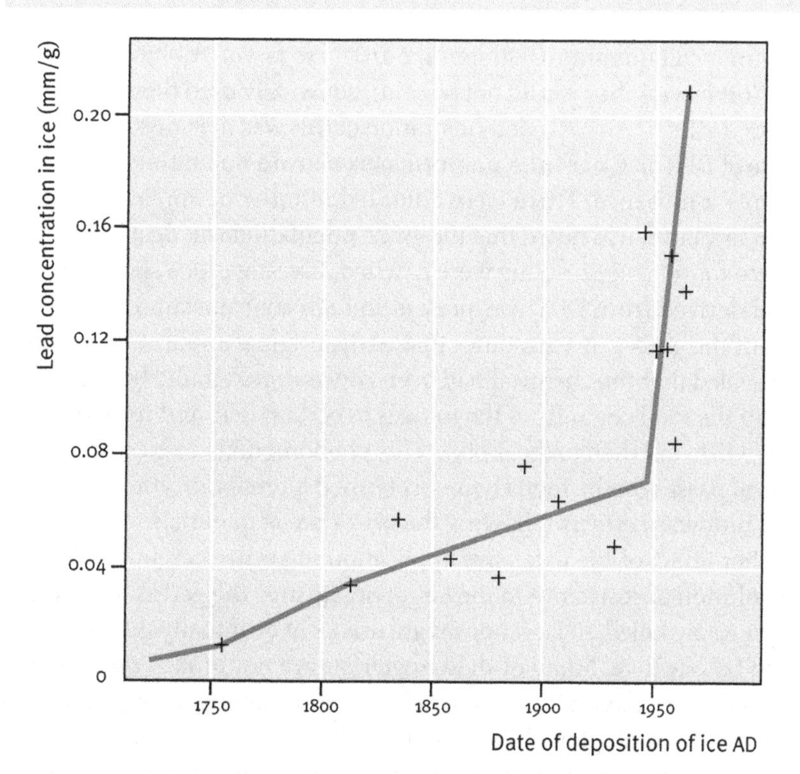

FIG. 12.8 The lead content of annual ice layers in Greenland (after Murozumi, Chow and Patterson, 1969)

An interesting case is the effect of lead on swans on the English rivers, particularly the Thames. Anglers used lead weights on their lines; they would sometimes lose these when the line became tangled with water plants and debris and swans would then swallow the weights, just as they swallow small pebbles to act as grind stones in their gizzards. Grinding up these small balls of lead the swans got a substantial dose of lead. Those that were suffering had a distinctive crick in their necks; frequently, these birds would die, but if they survived they would fail to raise young. The swan population was therefore falling. Then just over a decade ago anglers re-placed lead weights with those based on tungsten, and since that time the numbers of swans has steadily risen. This shows that what may appear to be a relatively minor pollutant may have a major effect on a member of the fauna.

A different class of pollutants are those compounds synthesized by the chemical industry; by and large they make a very positive contribution to our lifestyle. Some, like aspirin and many antibiotics, are present in nature but only in small quantities. Others, like DDT, are novel. All have apparent benefits (otherwise they would not be sold), but widely used they often have unintended effects. An early demonstration of this was the now classic story of the use of DDT in Clear Lake, California to control a non-biting midge that was simply a nuisance. There were a limited number of applications over several years, but it was noted that the grebe population was declining: dead birds were found and no young were hatched. Investigations showed that a chemical derived from DDT was present in high concentrations in the visceral fat of the grebes; 80,000 times that of the original application. Further work revealed that the chemical had been concentrated in the body fat as it passed up the food chain from the midges to certain fish, and from them to predatory fish and then to the grebes. This phenomenon—concentration as a chemical passes up the food chain—is termed biomagnification and causes many untoward effects following the use of novel materials.

Another effect of the extensive application of antibiotics and pesticides is the evolution of resistance in the target organisms: the germs or pests are no longer easily killed, so larger doses are used, but eventually the treatment ceases to be effective. Many of these materials are not broken down in the soil or water; they have a long residence time and, passing through the food chains, soon become globally distributed—DDT (or its product DDD) has now spread so far that it is found in polar bears in the north and in penguins in the south.

A further type of pollution arises when a material that is widespread and perhaps a vital part of the natural cycles that pass materials round the earth—biogeochemical cycles—becomes much more abundant through human activities. The outstanding example of this is carbon dioxide which is released when fossil fuels are burnt (FIG. 12.9). It is one of the key raw materials used in photosynthesis and therefore contributes to the base of all major food chains. When the molecules produced by photosynthesizing organisms (bacteria, algae and plants) are used as food by themselves and by organisms, ranging from bacteria to elephants, some carbon dioxide is released as they respire. This carbon dioxide passes out into the air or water and, in a balanced system, would be taken in again in photosynthesis. However, there is a great deal of carbon that has been 'stored' in the past: as peat or ooze, transformed by heat and pressure to fossil fuels (gas, oil, and coal) and as calcium carbonate in limestone and chalk, sometimes formed from the shells and skeletons of animals. When we burn fossil fuels—for example in cars and power stations—or make cement from chalk or lime-

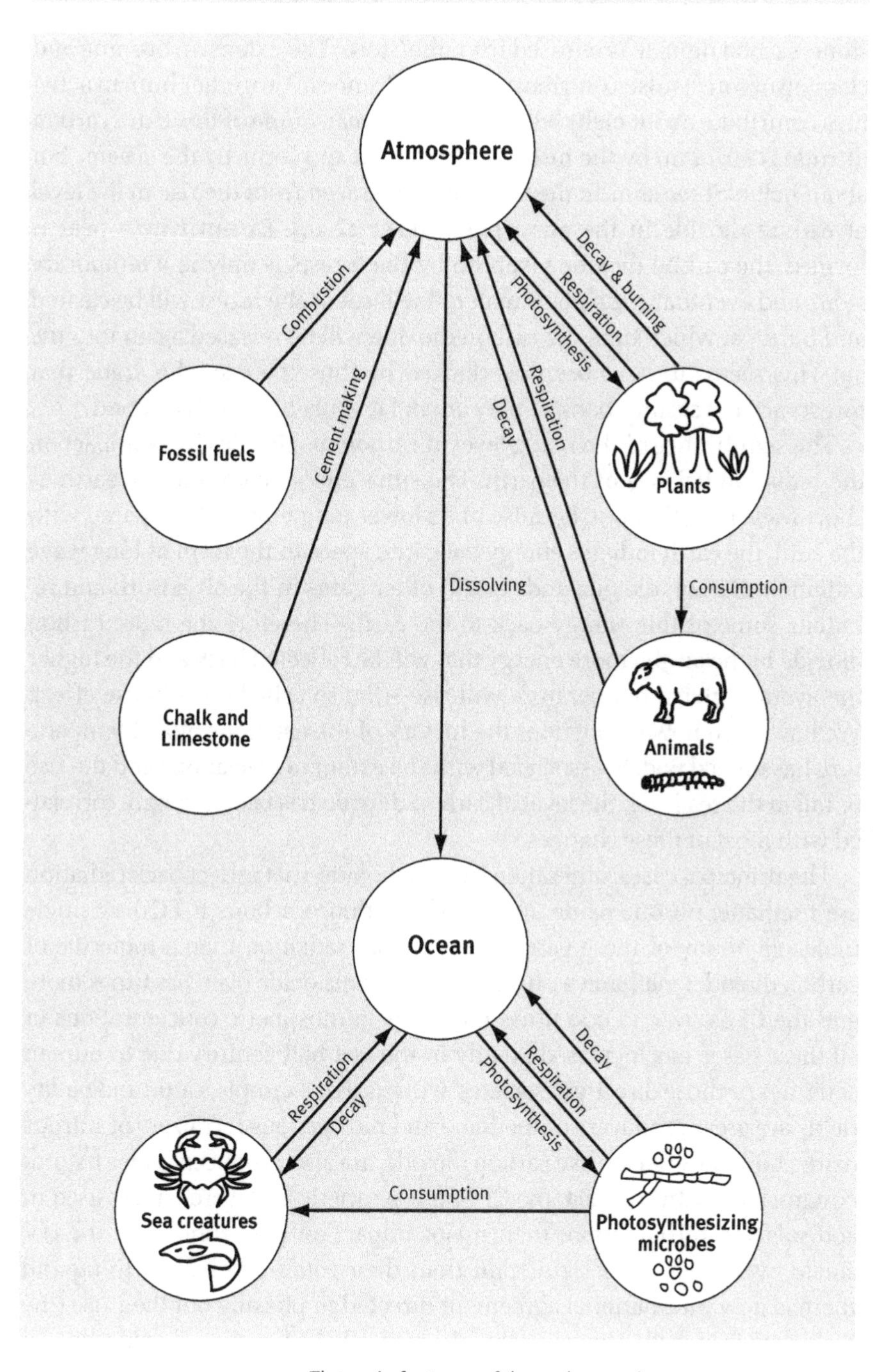

FIG. 12.9 The main features of the carbon cycle

stone, carbon dioxide is released from the 'store'. The extensive burning and clearing of forests also contributes, as do volcanoes. Altogether human activities contribute about eight billion tons each year. Some of this extra carbon dioxide is taken up by the forests of the world and some by the oceans, but about half of it remains in the air. This can be seen from the rise in the level of carbon dioxide in the atmosphere (FIG. 12.10). Except where peat is formed, the carbon dioxide taken up by the forests is only in a temporary store, and eventually the plant material will rot or the forest will be cleared and burnt, at which time the carbon dioxide will be released again into the air. This seems to have been overlooked by those people who argue that forests act as a secure ' bank'—they are in fact only a short-term 'bond'.

The significance of the rising level of carbon dioxide lies in its impact on the radiation balance of the earth. The sun's energy comes to the earth as short-wave radiation but, because of its lower temperature (compared with the sun), the earth radiates energy back into space in the form of long-wave radiation. Carbon dioxide and certain other gases in the air adsorb and re-radiate some of this energy back to the earth. Therefore the more carbon dioxide in the air the more energy that will be reflected back and the higher the average global temperature will rise—the so called greenhouse effect. We have seen how throughout the history of the earth the global temperature has ranged widely associated with the extent of glaciations and the rise or fall in the sea level; the level of carbon dioxide has been strongly correlated with most of these changes.

The principal gases, other than carbon dioxide, that reflect back radiation are methane, nitrous oxide and the chlorofluorocarbons (CFCs). A single molecule of any of these gases reflects more radiation than a molecule of carbon dioxide: methane 21 times more, nitrous oxide over 200 times more and the CFCs over 12,000 times more. The atmospheric concentrations of all these gases has increased greatly in the last half century due to human activities or those directly associated with us. For example, cattle and paddy fields are great producers of methane and nitrogenous fertilizers of nitrous oxide; both these gases, like carbon dioxide, are also produced under natural circumstances. In contrast, the CFCs are synthetic and were widely used in aerosols and refrigeration; their major impact on the ozone layer, the UV shield, was even more significant than their role in global warming and there is now international agreement directed to phasing out their use (the Montreal Protocol). However the elevated level of carbon dioxide has an effect about equivalent to twice that of these other greenhouse gases.

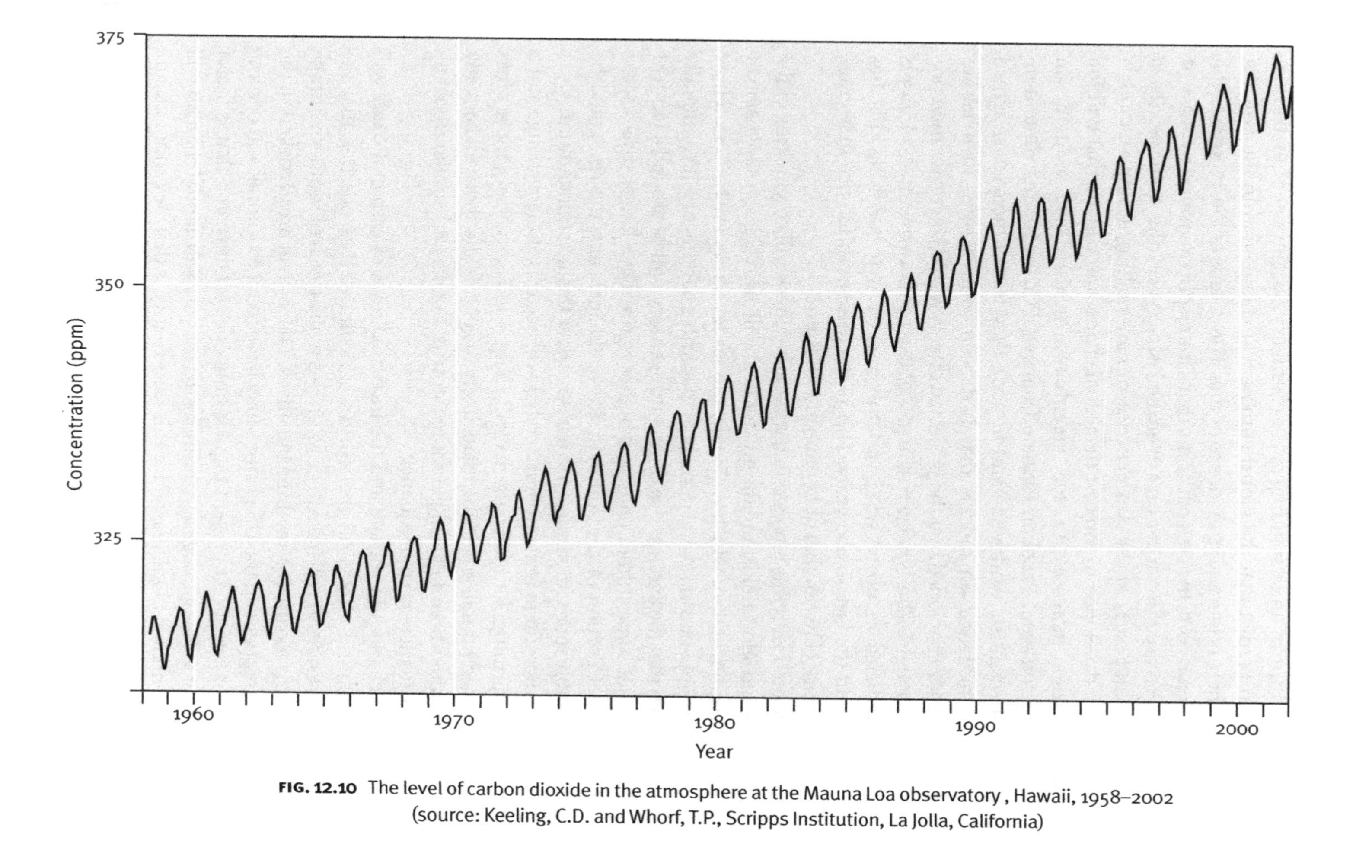

FIG. 12.10 The level of carbon dioxide in the atmosphere at the Mauna Loa observatory , Hawaii, 1958–2002 (source: Keeling, C.D. and Whorf, T.P., Scripps Institution, La Jolla, California)

Thus, from an understanding of the chemistry and physics of meteorology, we can predict that the global climate will change and in the last few decades there is increasing evidence of this. The decade of the 1990s was the warmest on record, the area of the northern hemisphere covered by snow or ice has decreased by 10 per cent over the last 40 years, the Arctic sea ice is thinning and parts of the Antarctic ice cap are melting. Complex models have been developed to forecast the actual changes in climate that will follow various increases in the greenhouse gases. There is now an Intergovernmental Panel on Climate Change (IPCC) involving some hundreds of scientists worldwide and in 2001 they issued their latest agreed conclusions based on models that had been refined since their earlier report. They concluded that the projected effects would be greater than previously thought. The average rise in global temperature over this century is now predicted as up to 5.8°C, certainly greater than anything that has occurred in the last 10,000 years. This is the present difference in average temperature between that of London and of Rome.

Associated with this increase in temperature there will be a melting of the ice caps and snow fields, and the greater heat will also cause an expansion of the sea water, both of which will make the sea level rise. We have seen how this has happened in several periods of the earth's past history. It is generally estimated that the average rise during this century will be around a tenth of a metre, but some studies point to a higher figure. On a longer time scale, a substantial portion of the Antarctic ice could melt and the rise could be considerably more, anything up to 6 metres over the next millennium.

It is important to remember that what will occur is climate change, not a benign warming. Rainfall patterns will change and, perversely, those areas with limited rainfall are, in the main, likely to get less and those that already have abundant rainfall will get more. Storms, floods, and similar disasters are likely to increase in frequency.

These changes in climate and the increased level of carbon dioxide will have profound effects on the world's agriculture. It is possible that the farmer in England will have to grow grapes while much of southern Europe will become too dry for anything but olives. Those great bread baskets of the Russian steppes and the US prairies may have reduced rainfall and become too dry for cereals. At present there is only one certainty, and that is that if the level of greenhouse gases continues to rise, very substantial changes will e forced upon us. One reason for this uncertainty is that we cannot be sure t all the changes will be gradual. There may be step changes, very sudden and unexpected changes. One possible trigger for a violent acceleration in change would be if the rise in sea temperature caused the vast amount of methane held in inert deposits in the sea to be released. Perhaps more likely is a change in the direction of ocean currents. Western Europe has been warmed for about the last 10,000 years by the Gulf Stream that flows up from the tropical Sargasso Sea. It is part of an ocean circulation that is driven, it is believed, by the cold salty water that occurs between Greenland and Norway. This is heavy (being cold and salty) and so it sinks and flows south in the North Atlantic generating a reverse surface current of warm tropical water (the Gulf Stream) moving north. The cold current would cease to sink if the water became less salty because of an increased input of freshwater from melting sea ice. There would be nothing to push the Gulf Stream and nowhere for it to go. The climate of London could become like that of Labrador—as it was 12,000 years ago. Evidence suggests that the Gulf Stream started quite quickly; perhaps its flow could stop over a few decades.

These impacts, and many more, arise as a direct result of human activities. They become more intense as the human population grows (FIG. 12.6), for example the more rice that has to be grown and the more cattle kept for food the more methane that is produced. But the overwhelming factor is the burning of fossil fuel to produce energy. The human impact (I) can be simply expressed in the equation:

$$I = P\,E\,N$$

where P = population, E = energy use (including that taken as food) per capita, and N = the proportion of activity that involves non-renewable resources, (resources like oil which when used will not be replaced by natural processes). Conversely, the sun and wind are renewable resources and we can use them to any extent to produce energy without depriving future generations; they are sustainable sources of power.

There are enormous variations in the average amount of energy used by individuals in different countries. In the USA the amount is twice that of the average European and 26 times that of an Indian. Unfortunately energy use is often seen as being coupled with economic growth and prosperity. Although this has been largely true in the last century, the scenario outlined above indicates that the two will have to be decoupled in the next century if we are not, as a world, to start to follow the path of the Easter Islanders. Will humans having made so much progress by increasing the carrying capacity of their habitat finally end by overexploiting the world and giving the kaleidoscope another shake? But life is flexible, and we can be sure that the frame of the kaleidoscope will be filled with a new pattern of colours. In contrast we, in our prodigious numbers, are locked by our agricultural and commercial activities into the current climatic regime. Can political stability survive the stresses that will arise when this changes or will we doom ourselves? We carry a burden of responsibility to learn from our knowledge of the world and its past. Time is short, but we do have the ability to change.

Further Reading

MANY of the topics discussed in this book are the subject of active research worldwide. Important new discoveries or interpretations are likely to be reported in *New Scientist* and *Scientific American*, and in a more technical form in *Nature* and *Science*.

General

Cowen, R. (1989) *History of life*. Blackwell Scientific Publications, Oxford.

Dawkins, R. (1995) *River out of Eden. A Darwinian view of life*. Weidenfeld & Nicholson, London.

Fortey, R. (1997) *Life: An unauthorised biography*. Harper Collins, London.

Kemp, T.S. (1999) *Fossils and evolution*. Oxford University Press.

Margulis, L. and Schwartz K.V. (1998) *Five kingdoms. An illustrated guide to the phyla of life on Earth*. W. H. Freeman, New York.

Schilthhuizen, M. (2001) *Frogs, flies, and dandelions: the making of species*. Oxford University Press.

Stanley, S.M. (1989) *Earth and life through time*. W. H. Freeman, New York.

Tudge, C. (2000) *The variety of life*. Oxford University Press.

Chapters 1–5

Conway Morris, S. (1998) *The crucible of creation*. Oxford University Press

Gould, S.J. (1991) *Wonderful life: the Burgess Shale and the nature of history*. Penguin, Harmondsworth.

Fortey, R . (2000) *Trilobite! Eyewitness to evolution*. Harper Collins, London.

Schopf, J.W. (1999) *Cradle of life*. Princeton University Press, Princeton, NJ.

Chapters 6–10

Behrensmeyer, A.K., Damuth, J.D. *et al.* (1992) *Terrestrial ecosystems through time*. Chicago University Press, Chicargo Ill.

Benton, M. and Harper, D. (1997) *Basic palaeontology*. Longman, Harlow.

Clarkson, E.N.K. (1998) *Invertebrate palaeontology and evolution*. Blackwell, Oxford.

Courtillot, V. (1999) *Evolutionary catastrophes*. Cambridge University Press.

Farlow, J.O. and Brett-Surman, M.K. (editors) (1999) *The complete dinosaur.* Indiana University Press.
Fastovsky, D.E. & Weishampel, D.B. (1996) *The evolution and extinction of the dinosaurs.* Cambridge University Press.
Frakes, L.A., Francis, J.E. & Syktus, J.I. (1992) *Climate modes of the phanerozoic.* Cambridge University Press.
Hallam, A. and Wignall, P.B. (1997) *Mass extinctions and their aftermath.* Oxford University Press.
McGowan, C. D. (1991) *Dinosaurs, spitfires, and sea dragons.* Harvard University Press, Cambridge, MA.
Morley, R.J. (2000) *The origin and evolution of tropical rain forests.* John Wiley and Sons, New York, NY.
Wilson, R.C.L., Drury, S.A., and Chapman, J.L. (2000) *The great ice age. Climate change and life.* Routledge, London.
Wood, R. (1999) *Reef evolution.* Oxford University Press.

Chapters 11–12

Byrne, R. (1995) *The thinking ape.* Oxford University Press.
Diamond, J. (1991) *The rise and fall of the third chimpanzee.* Vintage, London.
Ehrlich, P.R. (2000) *Human natures. genes, cultures and the human prospect.* Shearwater Books, Washington DC.
Ehrlich, P.R. and Ehrlich, A.H. (1972) *Population resources environment.* W. H. Freeman, New York.
Goudie, A. (2000) *The human impact on the natural environment.* Blackwell, Oxford.
Kingdon, J. (1993) *Self-made man and his undoing.* Simon & Schuster, London.
Klein, R.G. (1999) *The human career. Human biological and cultural origins.* Chicago University Press.
Tattersall, I. (1995) *The fossil trail.* Oxford University Press.
Wilson, E.O. (2000) *The future of life.* Little Brown, London.

Figure Acknowledgements

The author is grateful to the following for granting permission to redraw copyright material:

Dr L. C . Aiello, Fig. 11.3; Annals Missouri Botanical Gardens, Fig. 10.1; Blackwell Publishing, Fig.s 3.4, 5.3, 5.4, 5.5c,d, 5.6b, 5.7, 5.8, 5.9a,b, 5.10, 5.11b, 5.14b,d, 6.4a, 6.9, 7.6, 7.8, 7.12; Cambridge University Press, Fig. 6.14; Dr R. Carroll, Fig. 6.6; Dr M. A. Cluver (South African Museum), Fig. 7.10; Prof.s D. Dilcher & P. S. Crane, Fig 9.14; Prof. G. E. Fogg, Fig. 9.8b; Geological Society of America & University of Kansas Press, Fig. 8.8; Dr J. B. Graham, Fig. 7.1; IFREMER, Plouzané, France, Fig. 3.8; Drs C. D. Keeling & T. P. Whorf, Fig. 12.9; Dr T. Kemp, Fig.s 8.2, 8.4; Dr G King, Fig. 7.11c; Mr M. R. Long, Fig. 12.2d; Natural History Museum Trustees, Fig.s 3.9, 5.11a, 5.15, 8.5, 8.6c, 8.7, 9.1, 9.6, 10.14, 11.4, 12.5; Nature, Macmillan Magazines, Fig.s 6.6, 7, 9.15, 11.3; Dr E. Neumayer, Fig. 12.1; Oxford University Press, Australia, Fig. 10.8; National Research Council Canada, Research Press, Fig. 7.7; Palaeontological Association, Fig. 6.8a; Paleontological Society, Fig.s 5.1, 5.17; Pearson Education, Fig.s 5.5a,b; Mr J. Sibbick, Fig. 9.3; Prof R. A. Spicer Fig.s 6.12, 7.3, 7.5; Systematics Association, Fig.s 6.8b, 6.10, 6.11, 7.15, 7.16; Taylor & Francis Ltd., Fig. 9.11; Prof. H. B. Whittington, Fig. 5.6a; Dr C. Woese, Fig. 2.7; Dr J. Wolfe, Fig. 10.2a; Zoological Society of London, Fig. 5.16b.

Plate tectonic maps and continental drift animations by C. R. Scotese, PALEOMAP Project (www.scotese.com)

Index

NOTE: where a topic is continued on two or more successive pages only the first page number is given.

A

abominable snowman 216
Acheulean 224
acritarchs 29, 41, 62
Actinopterygians 71
adenosine triphosphate (ATP) 12, 23
aetosaurs 128
Agnatha 61, 68, 69
agriculturalist 239
Albertosaurus 146
algae 28, 41, 51, 55, 81, 114, 133, 158
Allegre, Claude 174
alligators 154
Alvarez, Luis 170
Alvarez, Walter 170
amber 134, 167
ammonia 75
ammonites 51, 77, 115, 130, 168
amphibians 97, 100
anaerobic habitats 23
Anaspids 101
Angiosperms 119, 137, 150, 163
Ankylosaurs *see* dinosaurs, armoured
Anomalocaris 56
antelopes 192
antibiotics 3, 250
ants 2, 78, 107, 184, 183
Apatosaurus 145, 147
apes 209, 213, 215
aphids 2
Archaean 7, 14
Archaeanthus 167
Archaebacteria 20, 28
archaeocyathids 46
Archaeopteris 85
Archaeopteryx 161
Aridipithecus ramidus 218
armadillo 235
artiodactyls 187, 189
asteroid 1, 8, 13, 16, 62, 87, 170
Atapuerca 224
atmosphere 7, 248
auks 202
Australopithecines 218
Australopithecus afarensis 219
autotrophs 10, 23

B

bacteria 17, 19, 33, 187
bacteriochlorophyll 17
bacteriorhodopsin 20
baleen 207
Baltica 66
banded iron formations 22
Barghoorn, Elso 17
barnacles 33, 78, 155
Baryonyx 140
Benton, Michael 128, 135
Bering land bridge 187, 197, 235, 242
biogeochemical cycles 250
biomagnification 250
biomineralization 45, 49
bipedalism 91, 104, 139, 217
birds 125, 161, 169, 181, 202, 235, 246
bison 192, 227, 235
bollide *see* asteroid
bony fish 68, 70
bovids 192
Brachiopoda *see* lampshells
Brachiosaurus 145, 149
brain/body ratio 222
bristle-cone pine 210
bristletail 99
bronze 246
buffalo 190
Burgess Shale 42, 56, 60
Byrne, Richard 223

C

Calamites 95
calcium carbonate 45
calcium phosphate 45
Cambrian 43
camels 191, 235, 242
Cann, Becky 230
capybara 192
carbon 9, 250
carbon cycle 91, 251
carbon dioxide 10, 74, 90, 117, 250, 252
carbon isotopes 14, 16, 116, 209
carboniferous 89
carrying capacity (K) 205, 233
cartilaginous fish 68, 110, 155, 168
caterpillars 105, 184
cats 194, 242
cattle 104, 192, 241, 242
caves, decorated 189, 230, 235
cell replication 12, 25
cells 11, 16, 25, 28
Cenozoic 176
centipedes 79
Cephalopoda 51, 68, 77, 130, 155
ceratopsians, *see* dinosaurs, horned
CFCs 252
Chaetoceros 157
chalk 156, 168, 252
Charniodiscus 40, 42
Châtelperronian artefacts 231
chemoautotrophs 10, 13, 22, 34
Chengjiang 56, 60
Chicxulub crater 172
chimpanzees 216
chlorophyll 17, 27
Choanozoa 30
Chordata 59, 79
chromosomes 16, 25
Chuaria 30
cities 246
civilization 246
cladistics 5
Cladoselache 71, 111
Clarkson, Euan 55
class 4
classification, biological system 4
claws 43
clay 12, 22
climate change 1
Cloud, Preston 17
Cloudina 41, 44, 51
Clubmosses 84, 137
Clutton-Brock, Tim 213
coal 89, 250
coccolithophorids 155, 168
coelacanth 72
coelom 31, 34
Coelophysis 128
co-evolution 163
Colinvaux, Colin 186
Collembola See spring tails
comets 8, 62, 135
competition 85, 178, 192
conifers 119, 137, 168
conodonts 55, 60, 68, 87, 135
consortium, 21, 50
continental drift 37; *see also* tectonic
convergence 52, 193, 199
Conway Morris, Simon 42, 55, 57
copper 246
corals 55, 68, 77, 87, 114, 133
cows *see* cattle

Cretaceous 136
crinoids *see* sea lilies
crocodiles 154, 169
Cro-Magnons 228, 234, 238
crops 240, 243
cucumber vine 183
Cyanobacteria 17, 19, 23, 28, 46, 54, 81
cycads 120, 137
cynodonts 122
cytoplasm 16

D
Darwin, Charles 2, 9, 228
Dawkins, Richard 2
DDT 250
Deccan Traps 137, 170
deer 192, 227
deforestation 115, 252
dehydration 74
determinate growth 94
deuterostomes 32
Devonian 65
Diamond, Jared 181
diaspids 101, 127
diatoms 156, 169
dicynodonts 102, 121
dingo 241
dinosaurs 77, 127, 136, 138, 146, 168, 191
dinosaurs and plants 149
dinosaurs, armoured 139, 142
dinosaurs, crested 141
dinosaurs, dome-headed 139, 141
dinosaurs, duck-billed 141, 148
dinosaurs, horned 139, 142
dinosaurs, plated 139, 142
diploblastic 31
Diplodocus 143 145, 149
diseases 243
DNA 2, 13, 15, 25, 216, 230
dodo 237
dogs 194, 235, 241
dolphin 207
domestication 234, 240
dragonflies 97, 99
drip tip 179
Dromaeosaurus 144
Dryopithecus 215
dwarfism 198

E
Earth 7
earthworms 75, 199
Easter Island 233, 245
echinoderms 53, 113
ecospace 53, 91, 234
Edaphosaurus 101
Ediacaran Fauna 39, 44
egg 27, 75, 99, 100, 111, 148, 153, 204
elephant birds 237
elephants 77, 126, 194, 227, 235, 242
Emeishan Traps 117
emperor penguin 204
Endosymbiotic Theory 27
energy 12, 32, 75, 255
energy used/capita 255
Eocene 176
epiphytes 78
Equisetales, *see* horsetails
Equus 188
Erects 216, 223
Eubacteria 20
Eukaryota 16, 27
Eurypterida, *see* sea scorpions
eutherians 193
eutrophic(ation) 47, 244
evolution 2, 15, 27, 32, 67, 74, 130, 163, 213, 225
extinction 62, 87, 114, 131, 134, 168, 189, 208, 235, 244

F
family 4, 62
farmer, *see* agriculturalist
feathers 161
ferns 84, 96, 119, 137, 169, 179
fertilizers 244
fire 235, 238
fire climax 239
fire-arms 238
fish 68
flight 109, 158
flowering plants, *see* angiosperms
food chains 23, 81, 156, 169, 183, 194, 250
foraminifers 114, 156, 168
Fortey, Richard 50
fossils 14
fungi 30, 82, 114, 174
fuse 32, 44, 47, 67, 128

G
galactic dust 11
gardens 246
gas 89, 250
Gastropoda 51
genes 2, 15, 47, 229, 231
genus 4
Giganotosaurus 146
Gigantopithecus 215
Gilbert, Larry 183
giraffes 192
glaciation 32, 62, 210, 252
global climate change 254
glucose 13, 23
Gnathostomulida 34
goats 192, 227, 242
Gondwana 39, 63, 65, 87, 136, 176
Goodman, Maurice 216
gorillas 216
Gould, Stephen J. 55, 225
graptolites 59
grass 186, 239
grasslands 186
Great American Interchange 177, 187
Great Leap Forward 233
greenhouse effect 32, 91, 115, 252
greenhouse gases 1, 117
Grypania 29
guillemot 203
Gulf Stream 255
Gunflint Chert 17
gymnosperms 119, 137, 150

H
hadrosaurs, *see* dinosaurs, duck-billed
halkieriids 57
Hallucigenia 56
handyman 220
haptomonads 155
Harvey, Paul 213
Heliconius butterfly 183
Helicoplacoids 53
Hennig, W. 5
herbivory 102
heterochrony 225
Heterodontosaurus 139
heterotrophs 10
hippopotami 126, 191
hoatzin 104, 182
hominins 209, 216
Homo
  antecessor 224
  erectus, see Erects
  ergaster 223
  habilis, see handyman
  neanderthalensis, see Neanderthals
  sapiens, see humans
horses 187, 227, 242
horsetails 84, 95, 137
human impact on the environment 255
humans 213, 218, 229, 233
hummingbirds 125, 183
hunter–gatherer 234
hydrothermal vents 11, 20, 33
Hypsilophodon 140

I
ice ages, see glaciations
ichthyosaurs 151, 168
Iguanodon 139
impact, human 255
Indricotherium 188
industry 246
insecticides 3
iridium 135, 170

iron 22, 45, 246, 248
iron pyrites 11, 22
irrigation 243
isotope 8, 14, 16, 116, 209

J
jacamar bird 183
jawless fish, *see* Agnatha
jellyfish 40, 68, 74
Jurassic 136

K
kangaroos 193, 235
Kemp, Tom 125
Kenyanthropus platyops 218
Kingdom 4, 82
Klein, Richard G. 224
koala bear 108
K-selected 205, 208, 237

L
labyrinthodonts 97
Laetoli footprints 217
lakes 248
Lambert, David 197
lampshells 51, 53, 68, 110, 114, 168
land, challenge of 74
language 229
Latimeria, see coelacanth
Laurasia 39, 63, 87, 118, 136, 176
lead 8, 246, 248
leaf-cutting ants 181
lemurs 213, 237
Lepidodendron 94
Lingula 53
Linnaeus Carl 4
litoptern 236
liverworts 30, 82
llamas 191, 242
lobefinned fish 71, 97
lobopods 57
lobsters 133
'Lucy' 219
lungfish 72
lycophytes, *see* clubmosses

M
Machiavellian intelligence 223
maidenhair tree 120, 137
mammals 121, 125, 169, 174, 235
mammoths 195, 212, 227, 235
Manicougan Crater 135
mankind, *see* humans
Manson Crater 172
manatees 208, 237
marsupials 193
mastodonts 197
mayflies 97, 99
Medullosa 96
meiofauna 35, 47
meiosis 26
membrane 12, 16, 25
mermaid's purse 111
Mesolithic 234
meteorites 8
Methanopyrus 33
Milankovich cycles 210
Millennium man 216
Miller, Stanley 10
millipedes 79
mimicry 185
minerals 234
Miocene 176, 215
mites 79
mitochondria 16, 27, 230
Mitochondrial Eve 230
mitosis 26
moas 237
mole rat 200
molecular clock 15, 169, 216
moles 199
molluscs 33, 51, 68, 87, 110, 168, 220
monkey puzzle 119, 137, 150
monkeys 182, 213
monomers 12
monophyletic 28
monoplacophorans 51
Montreal Protocol 252
Morgan, Elaine 226
mosasaurs 152, 168
mosses 30, 75
Mousterian 227
Mungo man 231, 235
murre 203
musk ox 190

N
natural selection 2
nautiloid 51, 68, 132, 169
Neanderland 227
Neanderthals 226, 233
nematodes 34
neoteny 225
newt 122
nitrogen 7, 9
nomenclature 4
non-renewable resources 255
notochord 59
nucleic acids 13
nucleus 16, 23
Nutcracker man 218
nuts 165, 218

O
Obik Sea 176
ocean currents 212, 255
oil 89, 155, 250, 255
old red sandstone 67
Oldowan 220
Olenidae 50
Oligocene 176
Oligokyphus 124
omomyids 213
Oparin, Aleksandr 9
orang pendek 224
order 4, 43
Ordovician 43
organelles 27
organic molecules 9, 12
ornithischians 138
Ornitholestes 143
ornithopods 139
Orrorin tugenensis 216
Ouranopithecus 216
Out of Africa migration 224, 228, 231
overexploited 244
Oviraptor 148
oxygen 9, 22, 23, 32, 50, 74, 90
oxygen isotopes 209
ozone 23, 74, 252

P
pachycephalosaurs, *see* dinosaurs, dome-headed
Palaeocene 176
Palaeolithic 221
Palaeo–Tethys Sea 89
panda 109
Panderichthys 73
Pangaea 38, 88, 89, 118, 136
Panthelassic Ocean 88, 118
Paranthropus 218
Parasaurolophus 144
Paratethys Sea 177
partridge, grey 243
passion vine 183
peat 250, 252
peccaries 191
pelycosaurs 101
penguin 203
perissodactyls 187
Permian 89
photosynthesis 13, 17, 77, 173, 250
phylogenetic tree 15, 21
phylogeny, 15, 28, 47
phylum 4, 79
phytosaurs 128
pigs 191, 242
Pilbara Range 17
placoderms 70, 87
plankton 32, 50, 77, 155, 173
plant defences 106
plastid 27
Pleistocene 176
plesiosaurs 152, 168
Pliocene 176, 217
pliosaurs 152
Polacanthus 140
pollen 166, 186
pollination 163
pollution 24, 246, 248
polymers 12

polyphyletic 28
Pompei worm 34
population, human 246, 255
primates 213
primordial soup 9, 13, 22
Problematica 55
Proconsul 215
progymnosperms 84, 95
Prokaryota 16
pronghorn 192
prosauropods 128
Proterozoic 25, 36
protists 29, 45, 114, 134, 155, 169
protoeukaryote 27
protostele 86
protostomes 32
Protozoa 29
Pteranodon 159
pteridophytes, *see* ferns
pteridosperms, *see* seed ferns
Pterobranchia 59
pterodactyloid 159
pterosaurs 158, 168
Pyrolobus 33

Q
Quaternary 176, 209

R
rabbits 104, 192
radiation 22, 116, 252
radioactivity 7, 9
radiolarians 114
rainfall 254
rainforest, tropical 180
rasps 43, 46
rays, *see* cartilaginous fish
redwoods 119
reefs 53, 87, 114, 133, 158
reproduction 27
reptiles, types of 101
resistance 3, 250
respiration 23, 74
rhinoceroses 126, 188, 212, 227, 234
rhyniophytes 82
Rickettsia 28
river blindness 248
RNA, 13, 22
rock paintings 234
rocks, oldest 16
rodents 92, 192, 200
Rodinia 38
rorquals 207
Rothamsted 244
Royal Palms 120
r-selected 205
rudists 158, 168
ruminants 187

S
saurischians 138
sauropods 145, 150, 195
schistosomiasis 248
Schoopf, William 22
scientific names 4
sea cows 209, 237
sea level 38, 47, 62, 88, 116, 135, 137, 170, 254
sea lilies 53, 68, 77, 113
sea scorpions 51, 68
seals 206
seaweeds 30, 45, 76, 79, 82
seed dispersal 164, 181
seed ferns 96, 99, 128
Seilacher, Adolph 42
self-weight 76, 92, 122, 189
Shark Bay 18, 87
sharks, *see* cartilaginous fish
sheep 192, 227, 240
shellfish, *see* molluscs
shells 46, 132, 168
shocked quartz 172
shrimps 112
Siberian Traps 115
silica 45
Silurian 65
siphonostele 86
Sivapithecus 215
skeleton 44, 76
skylark 244
sloth 182, 235
Snowball earth 33
solar system 1, 7
South Pole 64, 66, 87
species 4, 79
sphenophytes, *see* horsetails
spiders 78, 79
sponges 30, 55, 87, 114
sprawlers 122
Spray, John 135
Sprigg, Reginald C. 39
springtails 79
stegosaurs, *see* dinosaurs, plated
stonefly 110
stromatolites 17, 54
subfamily 4
sulphur 13, 20, 33, 35, 88
Sumerian 246
superfamily 4
swans 249
symbiosis 27, 103, 133, 192
synapsids 101, 121

T
tapirs 188
tarsier 214
taxa 4
tectonic movements and plates 8, 36, 87, 137
Teleostei 68; *see also* bony fish
temperature 78, 246, 252, 254
Tertiary 176
Tethys Ocean 118, 136, 176, 202
tetrapods 73, 97, 135
theory of mind 229
thermophiles 21
Therocephalia 121
theropods 128, 140, 144, 146, 150, 162
thorium 8
tools 220, 229, 234
tortoise 149, 169
Toxodon 236
trees 77, 85, 91, 104, 239
Triassic 118
Triceratops 144
trigonotarbids 79
trilobites 48, 68
trimerophytes 82
triploblastic 31
tropical rainforest 180
tuco-tuco 200
Turgai Strait 176
Turkana Boy 223
turtles 15 bn3, 220
Tyler, S.A. 17
Tyrannosaurus 139, 144, 146, 151

U
ungulates 187
uranium 8
urban 246
urea 76
Urey, Harold C. 10
uric acid 76

V
vendobionts 42
vertebrates 59
volcanoes 7, 62, 116, 174, 252
Volvulina 30
Vrba, Elizabeth 220

W
Wächtershäuser, Günter 10
wallabies 192
Wallace, Alfred 2
warm-blooded 125, 146
Warrawoona chert 17
water 7, 75, 217
water vapour 7
whales 76, 126, 207, 238
Whittington, Harry G. 56
wind 92
Wiwaxia 57
Wolfe, Jack 179
wombat, giant 235
woodlands 244
worms 33, 35, 41, 68, 75

Y
Younger Dryas period 212, 235

Z
zosterophylls 82
zygote 27